Vagus Nerve Stimulation and Anxiety

Taunjah P. Bell, Ph.D.

iUniverse, Inc.
New York   Bloomington

Vagus Nerve Stimulation and Anxiety

Copyright © 2010 by Taunjah P. Bell, Ph.D.

All rights reserved. No part of this book may be used or reproduced by any means, graphic, electronic, or mechanical, including photocopying, recording, taping or by any information storage retrieval system without the written permission of the publisher except in the case of brief quotations embodied in critical articles and reviews.

The views expressed in this work are solely those of the author and do not necessarily reflect the views of the publisher, and the publisher hereby disclaims any responsibility for them.

iUniverse books may be ordered through booksellers or by contacting:

iUniverse
1663 Liberty Drive
Bloomington, IN 47403
www.iuniverse.com
1-800-Authors (1-800-288-4677)

Because of the dynamic nature of the Internet, any Web addresses or links contained in this book may have changed since publication and may no longer be valid.

ISBN: 978-1-4502-4285-1 (sc)
ISBN: 978-1-4502-4286-8 (dj)
ISBN: 978-1-4502-4287-5 (ebk)

Printed in the United States of America

iUniverse rev. date: 7/16/2010

Photo Credits Page

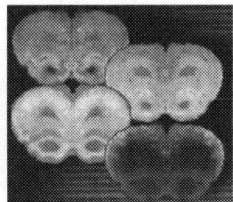

On the front cover is a montage of molecular images with a spectrum of peptides and proteins...their colors assigned to different molecular weights, in sections of a rat brain. Images like these, which were produced by MALDI imaging mass spectrometry, can reveal the distribution of individual proteins in cells and tissues. Dr. Malin Andersson and Dr. Richard Caprioli gave their permission for the author to reprint the montage of images of the rat brain that appeared in the winter 2007 issue of *Lens* magazine. Photo courtesy of the Vanderbilt Mass Spectrometry Research Center.

Taunjah P. Bell, Ph.D.

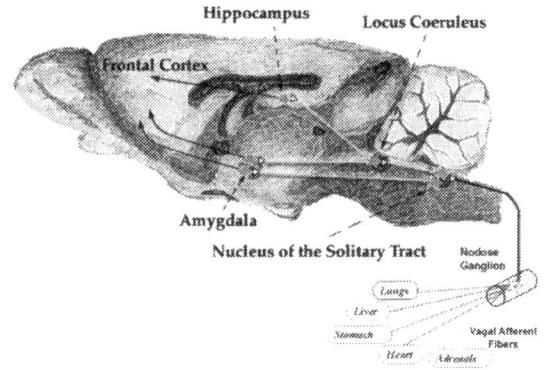

On the front cover is a diagram of a rat brain displaying vagal afferent fibers with cell bodies located in the nodose ganglion. These ascending sensory fibers relay information directly to the nucleus of the tractus solitarius (or NTS), commonly called the nucleus of the solitary tract. From the NTS, these afferent fibers make monosynaptic connections to the locus coeruleus (the main source of norepinephrine in the central nervous system) and disynaptic connections to the parabrachial nucleus. Photo courtesy of Dr. Cedric L. Williams, Associate Professor, Department of Psychology and Graduate Neuroscience Program, University of Virginia.

On the back cover is an image of a Long-Evans Hooded rat similar to the laboratory animals included in the research conducted by the author of this book. Photo courtesy of Taconic, Inc.

To my mother, Cleo Simmons Bell,
and
In memory of my father, Frank Bell, Sr.

Contents

Preface. xi
Acknowledgements xiii
Contributors xv
Chapter 1 Introduction 1
Chapter 2 Methods 9
Chapter 3 Results 31
Chapter 4 Discussion 47
Chapter 5 Conclusion 61
References 63
Index 75

PREFACE

The amygdala is the principle brain region implicated in the processing of fearful stimuli. In rats, distinct amygdalar pathways have been delineated for fear, a relatively brief, stimulus-specific response, and anxiety, a more sustained response prompted by conditions of unpredictable danger (Davis 1998; Davis, 2000; Davis, 2006). When anxiety becomes excessive and sustained in humans, it constitutes a disorder that is debilitating and difficult to treat. Previous research conducted in our laboratory suggested that in addition to enhancing memory processes, electrical stimulation of the vagus nerve in rats also attenuates anxiety-related behaviors. The purpose of the current research was to explore further the capacity of vagus nerve stimulation (VNS) to attenuate behaviors associated with anxiety. Another goal was to determine whether VNS (1.0 mA; 0.5 ms biphasic pulse width; 20 Hz; 30 sec train) produces its effects by activating descending parasympathetic efferent fibers. If a subcutaneous injection (0.1 mg/kg; 0.5 mg/kg; 1.0 mg/kg) of atropine methyl nitrate (AMN, a cholinergic muscarinic receptor antagonist that does not cross the blood-brain barrier) attenuated the effects of VNS, this finding would suggest that VNS acts by causing an increase in parasympathetic tone, specifically an increase in vagal tone.

One week before behavioral testing, 96 male Long-Evans rats were implanted with a bipolar stimulating electrode. After the recovery period, animals received two days of baseline testing on each of three measures of anxiety: open field, elevated plus-maze, and predator scent exposure task. Five minutes before each baseline session, the animal was placed in a shock box and a mild footshock (0.75 mA; 1.0 sec) was delivered to induce anxiety. On Day Three, 30 minutes prior to testing, each animal was administered either saline or AMN while unrestrained in its home cage. Twenty minutes later, VNS or sham stimulation was delivered via the head post while the animal moved freely in its home cage. Drug and stimulation effects were examined using a between-subjects multivariate analysis of variance (MANOVA) to analyze the mean scores of Day Three data. The results indicated that VNS had a robust and significant effect on all nine target behaviors measured. Atropine methyl nitrate at the high dose, and in some cases at the medium dose, had some capacity to attenuate the effects of VNS on most target behaviors. Rats administered AMN and no stimulation showed more anxiety than saline-treated sham-stimulated animals. These results provide strong support for the hypothesis that VNS attenuates anxiety.

ACKNOWLEDGEMENTS

I am truly thankful for the life that I have and eternally grateful for the opportunities that I have been given. I have been blessed in so many ways and have benefited from my academic and personal experiences on so many levels. Without my faith in God and the help of my family, significant other (Dr. Everett G. Neasman), dogs (Kate, Maya, Bo, and Jo), and so many special people in my life, I would not be where I am today. In particular, I must express my heartfelt appreciation to my significant other. Without Everett's love, advice, guidance, support, reassurance, and patience, my entire graduate and early professional career would have been even more difficult and quite unbearable. I will always be indebted to him for the sacrifices he has made to support my endeavors.

I also would like to extend my love, appreciation, and gratitude to my parents, Cleo Bell (mother) and Frank Bell, Sr. (father, deceased). I am truly grateful for the sacrifices they have made through the years so that I could achieve my academic, professional, and personal goals. Although my father is no longer with us, I am sure he would be very proud of my accomplishments. Their love, guidance, support, and commitment throughout the years are what have given me the strength and confidence to get where I am today. The journey has not been easy, but the experience has been worth everything I have endured. I know that the completion of this book is a milestone that my relatives, Everett's family (especially his parents Lydia J. Neasman and Farley B. Neasman, Sr.), our friends, neighbors, church members, mentors, and colleagues are proud of and I want them all to know they have been instrumental in my success.

Taunjah P. Bell, Ph.D.

In addition, I am genuinely appreciative of the valuable direction, compassionate advice, and professional expertise of my mentor Dr. Robert A. Jensen. I wish everyone had the opportunity to work with someone with such a passion for science and a zest for life. Dr. Jensen has contributed a great deal to ensure that this research was completed and my goals were accomplished. He has a genuine concern for others and his support has made all the difference in the world. I also would like to thank Dr. Jensen's neighbors' cat (Nemo) for donating the feline scent by wearing the cat collars used on Day Three of behavioral testing in the predator scent exposure task.

Furthermore, I extend my appreciation to Dr. Douglas C. Smith, Dr. Eric A. Jacobs, Dr. David G. Gilbert, and Dr. Thomas C. Calhoun for their individual expertise, recommendations, and guidance. Moreover, I would like to thank Dr. Smith for his involvement in this endeavor and the contribution of his laboratory space and equipment. I also would like to thank Dr. Jacobs for allowing me the opportunity to work in his laboratory during my first year at Southern Illinois University Carbondale and gain knowledge of handling animals, conducting behavioral testing, reviewing literature, and designing experiments that I would not have gained otherwise. In addition, I would like to thank Dr. Gilbert and Dr. Calhoun for their time, advice, careful reading, and dedication to this project. By providing guidance and offering professional advice, Dr. Calhoun who is currently the Chair of the Sociology Department at Jackson State University has been instrumental in my early career success as a new faculty member in the Psychology Department at Jackson State University.

I would like to express my gratitude to Cyberonics, Inc., for providing the Cyberonics Model 101 vagus nerve stimulators used in the present study and my appreciation to Pat A. McNeil (Associate Dean of the Graduate School), Dr. David L. Wilson (Dean and Director of the Graduate School), and the Graduate School Fellowship Office at Southern Illinois University Carbondale for granting me the James Walker Graduate Fellowship that provided the income I needed to pay tuition and cover household and living expenses during my last year of graduate school. Also, I am thankful for being awarded the James Walker Graduate Fellowship Research Award that funded the completion of my research.

CONTRIBUTORS

This contribution was supported by the James Walker Graduate Fellowship and the James Walker Graduate Fellowship Research Award granted to the author by Southern Illinois University Carbondale. Also, thanks are due to Cyberonics, Inc., for providing the vagus nerve stimulators that were used in the present study. In addition, the author is grateful for the academic support of Dr. Robert A. Jensen, Dr. Douglas C. Smith, Dr. David G. Gilbert, Dr. Thomas C. Calhoun, and Dr. Eric A. Jacobs.

1.
INTRODUCTION

For decades, researchers have attempted to understand the expression of emotional reactions and some of the underlying causes of anxiety disorders. Anxiety is a combination of psychological and cognitive states combined with a variety of somatic symptoms including heart palpitations, dry mouth, increased respiration, and gastric distress (Kandel, 2000). These symptoms appear to influence performance significantly in a variety of behavioral testing paradigms (Blatt & Takahashi, 1999; Ouagazzal, Kenny, & File, 1999) as well as in extralaboratory settings. Some neuroscientists have attributed the expression of anxiety-related symptoms to an overactive or over-reactive sympathetic nervous system (Carrasco & Van de Kar, 2003; Lin, Wang, & Shao, 2003; Orr, Metzger, & Pitman, 2002; Pohjavaara, Telaranta, & Vaisanen, 2003; Tancer, Lewis, & Stein, 1995).

However, substantial research evidence actually suggests that some symptoms of anxiety may be the result of a reduction in vagal tone resulting in a suppressed parasympathetic nervous system (Gershon, 1998; Levy, 1977; Porges, 1992, 1995a, 1995b, 1998, 2001; Prechtl & Powley, 1990). Vagal tone (i.e., activation of vagal fibers) is proposed as a novel index of stress and stress vulnerability in mammals. Porges' model emphasizes the role of the parasympathetic nervous system and particularly the vagus nerve in mediating homeostasis and defining stress. The model also emphasizes monitoring the neural control of the heart via the vagus (i.e., vagal tone) as an index of homeostasis.

Moreover, Porges' model details the importance of a branch of the vagus nerve originating in the nucleus ambiguus. By quantifying the amplitude of respiratory sinus arrhythmia with the proper training and adequate equipment, it is possible to assess the tonic and phasic regulation of the vagal pathways originating in the nucleus ambiguus.

In mammals, the nucleus ambiguus not only coordinates sucking, swallowing, and breathing, but it also regulates heart rate and vocalizations in response to stressful stimuli (Porges, 1992). Changes in the sucking, swallowing, breathing, heart rate, and vocalizations are usually representative of changes in vagal tone. Thus, these changes can serve as an indication of the level of stress in response to stressful stimuli or events experienced by the individual being monitored. Measurement of these components of vagal tone is proposed as a method to assess on an individual basis stress and stress vulnerability (Porges, 1995a). Therefore, in practical applications, this noninvasive method of monitoring vagal tone will allow the assessment of the stressful impact of various clinical treatments on patients and permit the identification of individuals with vulnerabilities to stress.

Previous research conducted in our laboratory suggests that specific components of the parasympathetic nervous system (i.e., the vagus nerve) have the capacity to attenuate anxiety-related behavioral responses in laboratory rats. The current research was designed to examine the role played by the parasympathetic nervous system (specifically the vagus nerve) in the expression of anxiety-related behaviors. This was accomplished through the induction of emotional arousal in rats and then by characterizing the effects of vagus nerve stimulation (VNS) on anxiety-related behaviors. Anti-anxiety (or anxiolytic) effects of VNS were assessed using the open field, elevated plus-maze, and the predator scent exposure task. These are well-established, reliable, and valid animal models of anxiety (Britton & Britton, 1981; File, 1985, 1992, 1997; File & Gonzalez, 1996; Hogg, 1996; Korte & De Boer, 2003; Prut & Belzung, 2003).

In the field of anxiety, assessment of behavior in the elevated plus-maze has become one of the most popular animal models. The elevated plus-maze is used extensively to identify anxiolytic and anxiogenic properties of experimental drugs (Pellow, Chopin, File, & Briley, 1985). Recently, there has been enormous growth in our understanding of the neurobiological basis of fear associated with anxiety. The bulk of this understanding is predicated on results of animal studies in which the capacity of experimental manipulations (e.g., drug injections, brain lesions, or microinfusions) to modify an animal's response to fear-inducing stimuli is evaluated. For example, when rats are confined to the open arms of the elevated plus-maze, they tend to exhibit significantly more fear-related behaviors and significantly higher plasma corticosteroid concentrations than when they are confined to the closed arms (Pellow, Chopin, File, & Briley, 1985). Behavioral testing using the elevated-plus maze involves placing a naïve animal (usually a rat or mouse) in the center of an elevated plus-maze with two open and two enclosed arms and allowing the animal to explore the maze freely during each behavioral testing session. Higher levels of locomotion and more time spent in the open arms of the maze is indicative of an animal showing lower levels of anxiety. Conversely, lower levels of locomotion and less time spent in the open arms of the maze is indicative of an animal exhibiting higher levels of anxiety (Pellow, Chopin, File, & Briley, 1985).

It has been suggested that the reluctance of rats to explore the open arms of the maze is associated with the fear of open spaces rather than the novelty or height of the maze (Fernandes & File, 1996; Pellow, Chopin, File, & Briley, 1985; Treit, Menard, & Royan, 1993). Moreover, anxiolytic compounds appear to increase the percentage of time rats spend on the open arms and the number of entries into the open arms, whereas anxiogenic compounds have the opposite effect (i.e., a decrease in the percentage of time spent on the open arm as well as the number of entries into the open arms) (Korte & De Boer, 2003; Pellow et al., 1985). The total number of entries and/or the number of entries into the closed arms reflects a measure of locomotor activity (Korte, De Boer, & Bohus, 1999).

Another popular method of assessing anxiety-like behavior in laboratory animals is the open field. Hall (1934) originally described the open field test for the study of emotionality in rats. Hall's apparatus consisted of a brightly illuminated circular arena approximately 1.20 m in diameter enclosed by walls approximately 0.45 m high. Hall's test involved placing a rat in the outer ring of the open field and observing the animal's behavior for two minutes. In some cases, rats were tested after 24 or 48 hours of food deprivation. Hall's researcher observed that rats walked more in the open field when they were food deprived. Some animals were labeled emotional and others were labeled non-emotional. Emotional animals made fewer entries into the central part of the arena. They also excreted greater amounts of urine and feces than the animals that were labeled as non-emotional animals (Hall, 1934).

The open field test is now one of the most popular procedures used in animal research on emotionality. Several versions are available, and they differ in shape of the environment (circular, rectangular, etc.), lighting (lighting from above or lighting from below where a bulb is placed under a translucent floor), color (sometimes a red light is used), or presence of objects (such as platforms, columns, tunnels, food, etc.) in the arena (Prut & Belzung, 2003). In contemporary research, the animal is typically placed in the center or close to the walls of the apparatus. The animal's behavior is usually recorded for a period ranging 2-20 minutes. During testing, researchers measure behaviors such as horizontal locomotion (number of crossings of the lines marked on the floor), frequency of rearing or leaning (sometimes termed vertical activity), and grooming.

Typically, rodents prefer to remain in the periphery of the open field rather than in the center of the apparatus. Mice and rats tend to walk close to the walls, a behavior called thigmotaxis. A high percentage of time spent in proximity to the walls, low levels of locomotion and an increase in freezing behavior are indicators of high levels of anxiety. Conversely, a low percentage of time spent in proximity to the walls, high levels of locomotion and a decrease in freezing behavior are indicators of low levels of anxiety (Prut & Belzung, 2003).

An influential movement in the study of anxiety has been a shift towards the use of animal models that have greater ethological relevance. This approach attempts to study anxiety by exposing laboratory animals to anxiogenic stimuli that closely mimic those seen in the wild rather than the more artificial situations that are often used in the laboratory (Rogers, 1997). Instead of exposing an animal to a maze, open field, footshock, or restraint, this approach might involve exposing a rat to a relevant predator (e.g., a cat) (Blanchard & Blanchard, 1972), a predatory cue (e.g., cat odor or owl call) (Blanchard, Blanchard, Weiss, & Meyer, 1990; Dielenberg, Arnold, & McGregor, 1999), or an alarm signal (e.g., ultrasonic vocalizations) (Beckett, Duxon, Aspley, Marsden, 1997; Brudzynski & Chiu, 1995). It can be argued that the use of such naturalistic stimuli allows for greater precision in the evocation of emotional states and defensive behaviors, thereby leading to improved modeling and analysis of fear and anxiety states.

While recent studies have focused on brain activation to anxiogenic stimuli, there are only a limited number of studies that have used more ethologically relevant stimuli. One example is a recent series of studies using *Fos* immunohistochemistry to identify the neural substrates brought into play when rats are exposed to a live cat (Canteras, Chiavegatto, Ribeiro Do Valle, & Swanson, 1997; Canteras & Goto, 1999). Compared to animals in the control condition, rats exposed to a cat exhibited a clear increase of *Fos*-like immunoreactivity in specific medial hypothalamic and periaqueductal gray (PAG) sites. These results differ from the results of previous studies involving the exposure of animals to an acute footshock that resulted in the expression of very little *Fos*-like immunoreactivity in the ventromedial hypothalamus and dorsal PAG (Li & Sawchenko, 1998; Pezzone, Lee, Hoffman, Pezzone, & Rabin, 1993). In contrast, acute footshock induces pronounced *Fos* expression in the nucleus of the tractus solitarius (NTS) and supraoptic nucleus as well as in the central, basolateral, and lateral amygdaloid nuclei (Li & Sawchenko, 1998; Pezzone, Lee, Hoffman, Pezzone, & Rabin, 1993), while exposure to a predator does not (Canteras, Chiavegatto, Ribeiro Do Valle, & Swanson, 1997; Canteras & Goto, 1999). These findings suggest that a distinct neural circuitry is activated as a result of predator-induced anxiety.

One study is especially relevant to our understanding of the involvement of the parasympathetic nervous system in the experience of fear, anxiety, and emotion. The goal of the study by Slaughter and Hahn (1974) was to examine the effects of electrical stimulation of the right vagus nerve on fear conditioning and subsequent avoidance behavior exhibited by rats. The rats were randomly assigned to one of five groups consisting of 10 subjects in each group. The groups were fear (F), no fear (NF), vagal stimulation (VS), vagal control (VC), and sympathetic stimulation (SS). All groups except the NF group received fear conditioning training with a tone followed by an inescapable shock.

The animals in the NF group underwent the same number of fear conditioning trials with a tone only.

Slaughter and Hahn (1974) implanted the animals with vagus nerve stimulating electrodes at the cervical level prior to their exposure to fear conditioning training. The VS group underwent fear conditioning training while simultaneously receiving VNS to decrease heart rate during those trials. Rats in four of the groups (i.e., F, VS, VC, and SS) learned to associate a tonal conditioned stimulus with inescapable shock. This resulted in conditioned freezing when the tone was subsequently presented. Such freezing behavior is inconsistent with behavior necessary for the acquisition of active avoidance responding in the presence of a tone. There was poorer avoidance and more freezing behavior seen in the rats that did experience an inescapable shock but did not receive VNS during fear conditioning trials compared to rats in the NF group that did not receive an inescapable shock during fear conditioning. Rats that received VNS during fear conditioning trials were able to perform the avoidance task significantly better than subjects that did not receive VNS during the fear conditioning trials. Vagus nerve stimulated animals showed far fewer incidents of freezing and avoidance behavior upon presentation of the conditioned stimulus than did rats in the three groups (i.e., F, VC, and SS) that did not receive VNS but did receive an inescapable shock. Slaughter and Hahn (1974) interpreted the results as indicating that stimulation of the vagus nerve produced a decrease in conditioned fear and anxiety resulting in better active avoidance performance.

The present study also was designed to determine whether or not atropine methyl nitrate (AMN), a peripherally-acting anxiogenic agent, had the capacity to attenuate the effects of VNS. Atropine methyl nitrate is a cholinergic muscarinic receptor antagonist that does not cross the blood-brain barrier (Shutt & Bowes, 1979). A successful cholinergic blockade of the anti-anxiety effects of VNS is a first step to understanding the mechanism of action of VNS on anxiety. Vagus nerve stimulation could exert its effects on anxiety either solely through actions in the central nervous system or by increasing peripheral parasympathetic nervous system tone thereby reducing some of the somatic components of anxiety. A cholinergic blockade of the peripheral parasympathetic system receptors on target organs tested this idea.

In the present study, it was hypothesized that left VNS delivered at the cervical level had the capacity to attenuate anxiety during behavioral testing using the three non-human animal models previously mentioned. Further, it was posited that VNS would induce its anxiolytic effects by activating descending parasympathetic nervous system efferent fibers and antagonism of the receptors on the target organs would be blocked by AMN thus reducing the effect of VNS. Moreover, it was proposed that the administration of AMN would attenuate the anti-anxiety effects of VNS in a dose-related fashion. Thus, it was expected that animals who received VNS and saline would exhibit reduced anxiety while subjects in the groups that received VNS and AMN would exhibit more anxiety.

2.
METHODS

2.1. ANIMALS

Ninety-six adult male Long-Evans rats obtained from Charles River Laboratories (Portage, Michigan) were used in this study. At the time of surgery, rats weighed approximately 300-350 g. Each animal was housed individually in a clear acrylic cage with wood chip bedding material. There was free access to food and water. Researchers as appropriate and Vivarium staff cared for the animals.

The rats were housed in the Vivarium where the colony room was maintained on a 12/12 h light/dark cycle with lights on 0600 to 1800 h and with air temperature of 21.0^0 C $\pm$ 2.0^0 (that is, between 19 and 23 degrees Celsius). This study was conducted in accordance with the ethical guidelines and regulations of the Southern Illinois University Carbondale Institutional Animal Care and Use Committee for experiments with laboratory animals. Protocol numbers 03-026 and 06-026 were approved for this study.

2.2. DESIGN AND CONSTRUCTION OF THE VAGUS NERVE STIMULATING ELECTRODES

The vagus nerve stimulating electrodes (see Figure 1) used in this study were constructed from two 7.0 mm sections of pure silver wire (A-M Systems, Inc.). First, each wire was soldered to a 30 ga, Kynar insulated, silver-plated stranded wire. Then, polyvinyl chloride (PVC) tubes (0.372 mm in diameter by 5.0 mm in length) were threaded onto each silver wire, and a strip of the PVC tubing (2.0 mm) was shaved from one side of each tube. This preparation exposed a segment of silver wire that made good contact with the vagus nerve.

Next, the two wires were inserted into a section of PVC tubing (3.5 mm) that served as a framework to keep the electrodes stable as well as separated 1.5 mm apart. The adhesive Epoxy (Walgreens) was used to insulate the end of each electrode pole. The seams joined the wire to the PVC tubing as well as filled the core of the framework tube. Black braided suture material (AliMed, Inc.) was attached to one electrode pole for later use to secure the electrode to the sternomastoid muscle.

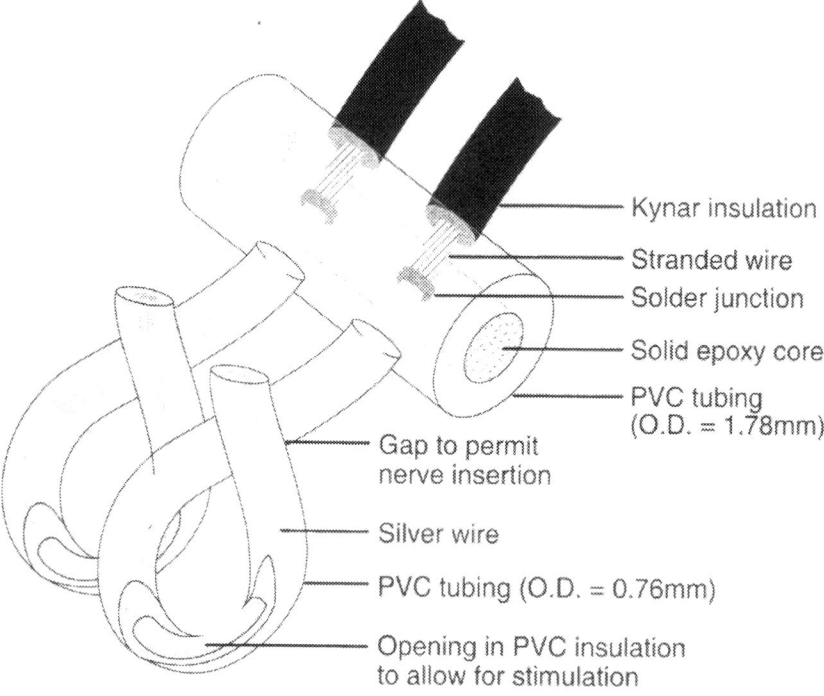

Figure 1. Stimulating electrodes were constructed from silver wire soldered to silver-plated stranded wire. PVC tubes were placed on the silver wires and a section was removed to expose a segment of silver wire that contacted the vagus nerve. Each stimulating electrode was formed into a helix that encircled the nerve when implanted. (From "A Study of the Effects of Vagus Nerve Stimulation on Anxiety in Laboratory Rats," by T. P. Bell, 2007, Unpublished doctoral dissertation, Southern Illinois University, Carbondale, Illinois.)

Then, each stimulating electrode was formed into a helix that encircled the vagus nerve (see Figure 2) when implanted. The helix formation allowed the nerve to make good electrical contact with the exposed segments of silver wire when the nerve was stimulated after VNS was delivered. The electrodes constructed in this study are consistent with the design of those used in the study by Smith, Modglin, Roosevelt, Neese, Jensen, Browning, et al. (2005).

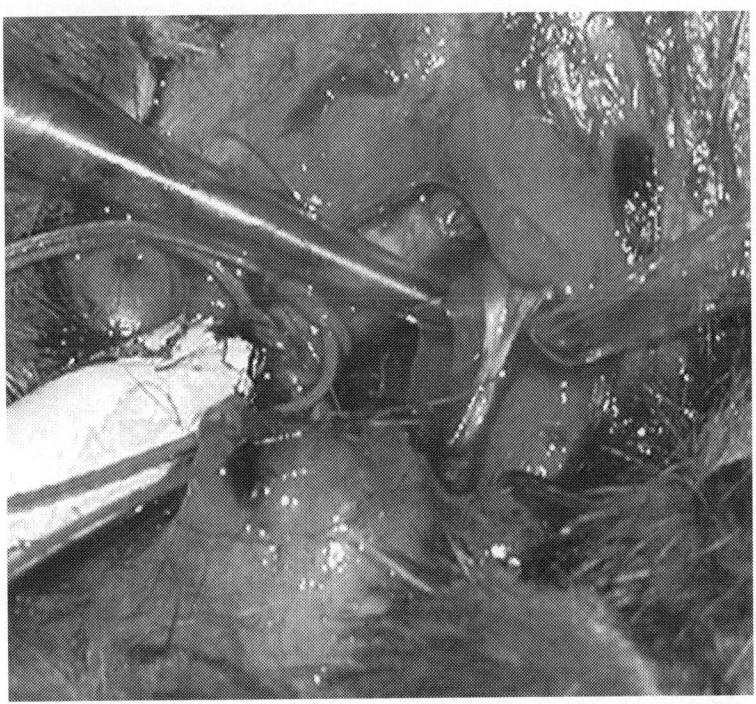

Figure 2. Photo of the left cervical vagus nerve of the rat.

2.3. VAGUS NERVE STIMULATING ELECTRODE IMPLANTATION SURGERY

Each animal was handled for five minutes everyday for two weeks before the surgery. Prior to the surgical procedure, each animal received an intraperitoneal (i.p.) injection of sodium pentobarbital (Nembutal) (65.0 mg/kg) (Henry Schein, Inc.) and a subcutaneous (s.c.) injection of atropine sulfate (0.04 mg/kg) (Sigma-Aldrich, Corporation). The atropine was administered to attenuate possible respiratory problems during and immediately following the surgery. Sterile surgical gloves were worn during all surgical procedures.

The surgical supplies and instruments were thoroughly cleaned in an ultrasonic cleanser (Harvard Apparatus) with antibacterial detergent (Walgreens) and sterilized in a surgical instrument sterilizer with hot glass beads (Fine Science Tools) that heat and cleanse instruments in 10-20 seconds at an operating temperature of approximately 250^0 C (i.e., 250 degrees Celsius). Instruments were allowed to cool to room temperature before use. All tools used during the surgical procedure were disinfected with 100% ethyl alcohol prior to and following surgery.

When each rat was successfully under anesthesia, the animal was shaved on the dorsal region of the head and the ventral region of the neck with electric clippers (Wahl Clipper, Corporation). The incision areas were swabbed with Povidone-iodine 10% solution (Walgreens). An incision approximately 2.5 cm in length was made on the ventral side of the neck just lateral to the midline to access the left vagus nerve at the cervical level. The layer of subcutaneous fat caudal to the salivary gland was blunt dissected to expose the intersection of the underlying sternomastoid and sternohyoid muscles. The muscles were longitudinally separated using small forceps and then the muscles were laterally retracted until the pulsing of the carotid artery could be clearly seen (Clark, Smith, Hassert, Browning, Naritoku, & Jensen, 1998).

Next, the vagus nerve was carefully separated from the surrounding connective tissue until a length of nerve sufficient for the placement of an electrode (ca. 7 mm) was exposed. The nerve was gently lifted and carefully placed into the bend of the helix of each pole of the stimulating electrode. After the vagus nerve was placed within each helix, the poles of the electrode were gently tightened to ensure adequate electrical contact with the vagus nerve. The stable framework tube supporting the electrode was sutured to the dorsomedial surface of the sternomastoid muscle to prevent displacement of the electrodes during head, neck, and trunk movements and locomotion.

The stranded wires from the electrode were threaded subcutaneously to a small incision made in the scalp just caudal to the skull and later attached to the head post that served as a connector to the vagus nerve stimulating electrode wires so that vagus nerve stimulation could be delivered through the head post via a Cyberonics, Inc. (Houston,

TX) NeuroCybernetic Prosthesis (NCP˚) System (Model 101). The incision on the ventral surface of the neck was closed with black braided suture material. Triple antibiotic ointment (Maximum Strength Triple Antibiotic Ointment Plus, Walgreens) was applied postoperatively to the incision area to prevent infection and to promote healing. Acetaminophen (Children's Tylenol, Cherry Flavor, Walgreens) was added to each animal's water bottle to alleviate pain and attenuate any swelling that might have occurred during or following surgery.

The animals were implanted with a stimulating electrode at the cervical level of the left vagus nerve because previous research conducted in our laboratory revealed that electrical stimulation of the right vagus nerve at the cervical level resulted in a high mortality rate. The researchers reported in laboratory notes that rats implanted with a right cervical vagus nerve stimulating electrode died at a higher rate than those implanted with a left cervical vagus nerve stimulating electrode. Because the right vagus nerve plays a greater role than the left vagus nerve in the control and regulation of heart rate, respiration, gastric tone, and other autonomic functions, the researchers noted that right vagus nerve stimulation at the cervical level interrupted normal autonomic functioning and tended to result in the death of the animals used in their study. Therefore, the animals used in the present study were implanted with a left vagus nerve stimulating electrode at the cervical level.

2.4. VAGUS NERVE STIMULATION HEAD POST PREPARATION

Immediately, following electrode placement, each rat was placed in a stereotaxic frame (Kopf Model 900), and a midline incision was made on the dorsal region of the head. This incision joined the dorsal incision made during the vagus nerve stimulating electrode implantation where the electrode wires emerged. The scalp was reflected so that the calvaria could be freed of the periosteum. Four 0.9 mm holes were drilled in the skull. Two holes were positioned just anterior to bregma and the other two holes were positioned approximately 2.5 mm posterior to bregma. Each hole was approximately 1.5 mm lateral to the midline. A stainless steel anchor screw (4.8 mm in length) (Plastics One, Inc.) was implanted in each hole.

Taunjah P. Bell, Ph.D.

A head post was constructed from a four-hole section of Amphenol Microminiature Strip Connector (Allied Electronics, Inc.) material. The wires from the vagus nerve stimulating electrode were cut to length, and gold-plated male connector pins (Allied Electronics, Inc.) were soldered to the wires. The pins were then inserted into the center two holes of the strip connector. The head post was positioned between the screws, and then it was attached to the skull and screws with dental acrylic cement (Co-Oral-Ite Dental Mfg.). The incision area was then carefully cleaned and closed with wound clips. The entire surgical procedure lasted approximately 1.25 hours.

Immediately following completion of the surgery, an intramuscular (i.m.) injection of penicillin (30,000 i.u.) (Henry Schein, Inc.) was administered to each animal. The injection site was massaged briefly to ensure adequate circulation of the drug. Triple antibiotic ointment was applied to the incision area to prevent infection and to promote healing. Each animal was monitored until it recovered from the anesthesia and locomoted normally. Each rat was then monitored daily (including weekends) during the seven-day recovery period to ensure proper healing of both incisions and sufficient maintenance of presurgery body weight (Markus, 2002).

2.5. ELECTRODE IMPEDANCE TESTING

The Cyberonics Model 250 NeuroCybernetic Prosthesis (NCP˚) Programming Software was used to program the NCP Pulse Generator and to assess the impedance level of each vagus nerve stimulating electrode. The impedance of each electrode was assessed following the completion of behavioral testing and prior to the euthanization of each animal tested and included in the present study. When testing the impedance of each electrode, the head post of each animal was connected to the NCP Pulse Generator via a four-hole section of Amphenol Microminiature Strip Connector with two gold-plated female pins connected to the two holes in the middle of the strip connector. The programming wand was placed over the NCP Pulse Generator allowing the software to deliver a 1.0 mA; 0.5 ms biphasic pulse width; 20 Hz; 30 sec train. These were the same parameters used for diagnostic data collection in the present study and in the previous study by Markus (2002). Following data collection, the impedance of the electrode was assigned a numerical value (DC-DC converter code) corresponding to a range of impedance values.

Prior to the start of this study, it was agreed upon that animals whose stimulating electrodes impedance value exceeded a DC-DC converter code of three would be excluded from the study and any data collected during the behavioral testing sessions would not be included in the statistical analyses. Fortunately, the data collected during all the behavioral testing sessions were included in the final statistical analyses because the estimated lead impedance of the vagus nerve stimulating electrodes implanted in the animals included in the present study did not exceed a DC-DC converter code of three (see Table 1).

TABLE 1
ESTIMATED LEAD IMPEDANCE

DC-DC Converter Code	Impedance Range	Status Reading
0	<1.00 K ohm	OK
1	1.01-3.00 K ohm	OK
2	3.01-5.00 K ohm	OK
3	5.01-8.00 K ohm	OK
4	8.01-11.00 K ohm	High
5	11.01-16.00 K ohm	High
6	16.01-20.00 K ohm	High
7	>20.01 K ohm	High

Note. OK = the impedance of the vagus nerve stimulating electrode is within an acceptable operating range and the data collected during the animal's behavioral testing sessions were included in the final statistical analyses; High = the impedance of the vagus nerve stimulating electrode is higher than expected and the data collected during the animal's behavioral testing sessions were not included in the final statistical analyses.

Taunjah P. Bell, Ph.D.

2.6. PRE-TEST LOCOMOTOR RESPONSE RATINGS

Seven days following the left cervical vagus nerve stimulating electrode implantation surgery described earlier, each animal was assigned to drug conditions based on pre-test activity scores. Prior to behavioral testing, each animal was exposed to the open field and elevated plus-maze for 300 seconds (five minutes). During this period, animals were videotaped and their locomotor activity was timed using a stopwatch and the results were recorded on paper. Each animal's open field and elevated plus-maze locomotor response was rated as low (0-1.66 minutes of locomotion), medium (1.67-3.33 minutes of locomotion), or high (3.34-5.0 minutes of locomotion) based on the amount of time spent exploring the open field or elevated plus-maze. Based on each rat's locomotor response rating, the animal was classified as either a low, medium, or high responder (LR, MR, or HR, respectively).

In the present study, we extended the system used by Gulley, Hoover, Larson, and Zahniser (2003) who investigated several factors that might contribute to cocaine-induced behavioral variability and its association with differential inhibition of dopaminergic transporter function. The researchers found that out-bred Sprague-Dawley rats can be classified as either low or high cocaine responders (LCRs or HCRs, respectively) based on their open field locomotor response to acute cocaine (10.0 mg/kg, i.p.). In a similar study, Briegleb, Gulley, Hoover, and Zahniser (2004) extended their analysis to amphetamine (0.5, 1.0, and 5.0 mg/kg, i.p.) and found that the individual differences in drug-induced behavioral activation and its association with differential inhibition of dopaminergic transporter function were not as pronounced as in the cocaine study.

2.7. DRUG ADMINISTRATION AND STIMULATION PRESENTATION

As previously mentioned, the drug condition to which each rat was exposed was based on the animal's pre-test locomotor response ratings. Therefore, animals were assigned to groups using stratified random assignment to ensure that there was an equal distribution of low, medium, and high responders across treatment conditions. In this study, there were four treatment conditions with 24 animals in each condition. Stimulation presentation to each animal in each group was randomly determined. Animals were surgically implanted with left vagus nerve stimulating electrodes that were attached to a head post that was prepared and securely mounted to the skull so that VNS could be delivered on Day Three of behavioral testing.

Seven days following surgery, subjects began a 12-day behavioral testing regimen (three days open field testing; three days elevated plus-maze testing; and three days predator scent exposure task testing with a one-day break between each test). Thirty minutes prior to behavioral testing on Day Three, animals received a subcutaneous (s.c.) injection of 0.9% sodium chloride (saline, 0.5 ml/kg) or atropine methyl nitrate (AMN) in one of three doses: 0.1 mg/kg (low dose), 0.5 mg/kg (medium dose) or 1.0 mg/kg (high dose) (Croiset, Raats, Nijsen, & Wiegant, 1994). According to Shutt and Bowes (1979), larger doses than 1.0 mg/kg of AMN should not cause a greater effect because the receptors might be saturated at doses of 1.0 mg/kg or higher. For example, doses of AMN higher than 2.0 mg/kg may cause the opposite or reverse of the intended effect (i.e., bradycardia, a slowing of the heart rate, instead of tachycardia, an increase in the heart rate; Shutt & Bowes, 1979). Moreover, in the circulation, approximately 50% of AMN is bound to the plasma proteins, and the plasma half-life of the drug is 2.5 hours. It has been shown that AMN, a cholinergic muscarinic antagonist and quaternary ammonium derivative, does not penetrate the blood-brain barrier because of its ionization properties (Shutt & Bowes, 1979).

Also, on Day Three, animals received VNS (1.0 mA or sham) 10 minutes prior to behavioral testing. Stimulation was delivered via the head post preparation while the animal was unrestrained and freely moving in its home cage. Therefore, 30 minutes prior to behavioral testing on Day Three, each rat received a subcutaneous (s.c.) injection of saline or AMN. The injection was administered while the animal was unrestrained and freely moving in its home cage. Then, 20 minutes later (that is, 10 minutes prior to testing), VNS or sham VNS was administered before each animal was placed in the behavioral testing apparatus.

Thus, on Day Three, animals received a subcutaneous injection of saline (0.5 ml/kg) or AMN (0.1 mg/kg; 0.5 mg/kg; or 1.0 mg/kg) 30 minutes before behavioral testing in the open field, elevated plus-maze, or predator scent exposure task. Therefore, on Day Three, prior to behavioral testing in the open field, elevated plus-maze, and predator scent exposure task, 12 subjects received VNS and saline (0.5 ml/kg, s.c.); 12 subjects received VNS and a low dose of AMN (0.1 mg/kg, s.c.); 12 subjects received VNS and a medium dose of AMN (0.5 mg/kg, s.c.); 12 subjects received VNS and a high dose of AMN (1.0 mg/kg, s.c.); 12 subjects received sham VNS and saline (0.5 ml/kg, s.c.); 12 subjects received sham VNS and a low dose of AMN (0.1 mg/kg, s.c.); 12 subjects received sham VNS and a medium dose of AMN (0.5 mg/kg, s.c.); and 12 subjects received sham VNS and a high dose of AMN (1.0 mg/kg, s.c.). The period of time between onset of each drug administration and end of each behavioral test was approximately 45 minutes. Each animal's performance on all behavioral tasks were recorded on videotape and scored on data collection sheets. Finally, the effects of drug administration and stimulation presentation were assessed during data analysis.

2.8. BEHAVIORAL TESTING PROCEDURE

All animals were handled for five minutes everyday for two weeks prior to surgery to ensure that they were manageable and non-aggressive (i.e., rats did not attempt to bite, scratch, or escape when the cage top was removed) during drug administration, stimulation presentation, and behavioral testing sessions. Half the animals received VNS (1.0 mA; 0.5 ms biphasic pulse width; 20 Hz; 30 sec train) and the other half received sham VNS. The order of testing for each animal in two (i.e., open field and elevated plus-maze) of the three behavioral tests was determined by the animal's pre-test locomotor response rating so that there was an equal distribution of low, medium, and high responders in each group as previously stated. Further, approximately half the animals received each of the two assessments of anxiety first, with the predator scent exposure task administered last across all conditions. The predator scent exposure task was administered last to ensure that the effects of cat odor exposure would not interfere with each animal's responses during behavioral testing in the other measures thereby eliminating the possibility of any carryover effects.

As mentioned earlier, seven days following the surgery previously described, animals began a 12-day testing regimen (three days open field testing, three days elevated plus-maze testing, and three days predator scent exposure task testing with a one-day break between each test). There was a one-day break between each measure to ensure that the dose of AMN administered on Day Three had been metabolized and eliminated from the animal's body. The major route of drug elimination from the body is renal (kidney) excretion of drug metabolites. Thus, most drugs leave the body in urine following the hepatic (liver) biodegradation of the drug into drug metabolites.

On Days One and Two prior to behavioral testing in the open field and elevated plus-maze, each subject was given a mild electric footshock (0.75 mA; 1.0 sec) in a separate enclosure (45.3 cm x 40.8 cm x 52.8 cm). This enclosure is made of wood, painted gray, and fitted with a grid floor made of 22 stainless steel rods (40.8 cm in length x 0.5 cm in diameter) spaced 1.5 cm apart. The lid (45.3 cm in length x 40.8 cm in diameter) of the enclosure is operated manually by lifting. Five minutes after delivery of the footshock, each animal was placed in either the open field or elevated plus-maze for behavioral testing. Administration of the footshock was used to induce a degree of experimental anxiety (Clark, Krahl, Smith, & Jensen, 1995).

On Day Three, half the subjects received sham VNS while the other half received VNS. Sham VNS or VNS was delivered while the animal moved freely in its home cage. This procedure was performed by carefully attaching a small plug to the previously implanted head post. The plug was made from a four-hole length Amphenol Microminiature Strip Connector material. Whether a subject received sham VNS or VNS was determined randomly. Following delivery of either sham VNS or VNS, the plug was gently removed from the animal's head post. Ten minutes later, each animal was placed in the open field, elevated plus-maze, or predator scent exposure task, and the subject's behavior was scored as previously described.

During the initial predator scent exposure task trials, the hide box and a control (unscented) cat collar suspended from a wooden dowel were present. There was no stimulation delivered during these initial sessions on Days One and Two. The predator-scented cat collar was introduced only on the last day (Day Three) of behavioral testing following delivery of sham VNS or VNS (Markus, 2002). The effects of the mild electric footshock, vagus nerve stimulation, and the following drug administration (saline or atropine methyl nitrate, AMN) on each animal's behavior were assessed. After the completion of all behavioral testing and prior to the euthanization of each experimental subject, the electrode impedance was checked to verify whether or not the silver wire made good contact with the vagus nerve. Animals were humanely euthanized by the Vivarium staff.

2.9. BEHAVIORAL TESTING APPARATUS

2.9.1. OPEN FIELD

A large square enclosure (100 cm x 100 cm) with walls 30 cm high was used as an open field (see Figure 3). The inside of the open field is painted gray. The floor of the open field is painted with white lines delineating sixteen 25 cm x 25 cm sectors. The same behavioral testing apparatus described here, the same behavioral responses measured and listed below, the same three target behaviors analyzed and mentioned later, and the same behavioral testing procedures outlined below were used in the unpublished doctoral dissertation written by Markus (2002).

Figure 3. Photo of the apparatus used during behavioral testing in the open field.

To begin behavioral testing in the present study, each animal was placed in the center of the open field. The frequency of rearing or leaning (vertical activity), number of lines crossed (horizontal locomotion), number of fecal boli, number of grooming bouts, number of entries into the four central sectors, time spent in the four central sectors, time spent in the perimeter sectors, time spent against the walls (thigmotaxis), and time spent freezing were assessed for 900 seconds (15 minutes) (Prut & Belzung, 2003). During the actual 15-minute behavioral testing session while the animal was in the open field, the frequency of rearing or leaning, number of grooming bouts, number of lines crossed, and number of entries into the four central sectors were assessed using a series of manually controlled counters.

Next, while viewing the videotape of the 15-minute behavioral testing session and using four manually controlled stopwatches (Radio Shack, Tandy Corporation), the researcher assessed time spent in the four central sectors, time spent in the perimeter sectors, time spent against the walls, and time spent freezing. At the end of each session, the number of fecal boli was counted via direct observation by the researcher who conducted the behavioral testing session. Therefore, during the 15-minute sessions, each animal's behavior was observed by the researcher, documented on a data collection sheet, and recorded on videotape. Also, each 15-minute behavioral testing period was timed using a battery-operated timer (Radio Shack, Tandy Corporation). Between trials, the open field was swabbed with a 20% ethanol solution to attenuate odor cues.

The above behaviors were selected for observation during behavioral testing because they are associated with the induction of anxiety. According to the literature, frequent rearing and leaning is exhibited by rats that appear to be experiencing anxiety, fear, or stress following exposure to certain types of stimuli (e.g., predators, loud noises, bright lights, etc.). Animals that remain in the same place (e.g., do not cross many or any lines in the open field, do not enter the center of the field, hide in the corners, etc.), prefer the perimeter and cling to the walls, or engage in freezing behavior for long periods of time during behavioral testing are believed to be experiencing anxiety or fear (Power & McGaugh, 2002; Vazdarjanova, Cahill, & McGaugh, 2001).

Also, urination and the excretion of fecal boli (defecation), especially in large quantities, are interpreted as the exhibition of anxiety-related behavioral responses (Hall, 1934). In many cases, rats that are not well-groomed and rats that fail to groom themselves regularly or frequently are usually monitored by experimenters, veterinarians, or caretakers for signs of illness. Because rats usually groom themselves frequently, one that fails to maintain a neat and clean appearance is believed to be experiencing some type of physical or psychological crisis. Similarly, rats that do not frequently engage in grooming bouts during the behavioral testing sessions are believed to be experiencing anxiety, especially those that have been exposed to anxiety-inducing stimuli, such as loud noises, bright lights, a mild electric footshock, or AMN, an anxiogenic agent. Moreover, the open field was selected for use in the present study because in experiments testing animal models of anxiety, it has proven to be a reliable and credible behavioral measure according to the literature. The three target behaviors (number of sectors crossed in the open field, number of entries into the four central sectors, and time spent in the four central sectors) selected for statistical analyses exemplify the epitome of anxiety-related behavioral responses motivated by fear-arousing stimuli.

During the behavioral testing sessions in the present study, some behavioral responses were counted, other behavioral responses were timed, and all target behaviors were documented, scored and recorded on videotape. Later, each animal's behavior as recorded was scored by a second observer who was blind to the treatment condition to which the animal was exposed (Britton & Britton, 1981; Markus, 2002). In other words, the second observer was unaware of the treatment condition to which the animal was exposed. The data from both observers was used to establish inter-rater reliability.

2.9.2. ELEVATED PLUS-MAZE

The elevated plus-maze is colored gray with three white lines approximately 15 cm apart painted only on the open arms (see Figure 4). This apparatus consists of two open arms (50 cm long x 10 cm wide) and two enclosed arms (40 cm long x 10 cm wide). The walls of the apparatus are 40 cm high, and the four arms extend from a central platform (10 cm x 10 cm). The maze is supported by four legs so that the walkways are 50 cm above the floor. Each of the open arms is divided into three equal sectors. A mirror was affixed to the top of one closed arm to facilitate observation of behavior occurring in the opposite closed arm from across the room. The same behavioral testing apparatus described here, the same behavioral responses recorded and measured, the same target behaviors analyzed, and the same behavioral testing procedures employed in this study were used in previous research by Markus (2002). Time spent in the central platform, time spent in the open arms, time spent in the closed arms, time spent against the walls, time spent freezing, number of sectors crossed in the open arms, number of entries in the central sector, number of leans, number of rears, number of fecal boli, and the number of grooming bouts were counted using a series of counters and timers controlled manually (Markus, 2002). In addition, thigmotaxis and vertical activity were recorded during the 15-minute test period (Prut & Belzung, 2003).

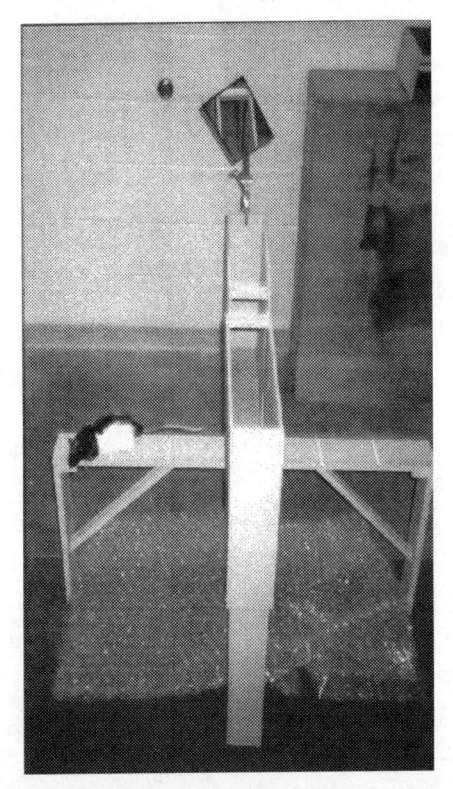

Figure 4. Photo of the apparatus used during behavioral testing in the elevated plus-maze.

To begin behavioral testing in the present study, each animal was placed in the center of the elevated plus-maze on the central platform. During the actual 15-minute behavioral testing session while the animal was in the maze, the number of rears, number of leans, number of grooming bouts, number of entries into the central sector, and number of sectors crossed in the open arms were assessed using a series of manually controlled counters. Next, while viewing the videotape of the 15-minute behavioral testing session and using four manually controlled stopwatches, the researcher assessed time spent on the central platform, time spent in the open arms, time spent against the walls, and time spent freezing. Time spent in the closed arms of the elevated plus-maze was calculated by summing time spent on the central platform and time spent in the open arms and subtracting the total from 15 minutes (the entire length of time allotted for each behavioral testing session). At the end of each session, the number of fecal boli was counted via direct observation by the researcher who conducted the behavioral testing session. Therefore, during the 15-minute sessions, each animal's behavior was observed by the researcher, documented on a data collection sheet, and recorded on videotape. In addition, each 15-minute behavioral testing period was timed using a battery-operated timer. Between trials, the elevated plus-maze was swabbed with a 20% ethanol solution to attenuate odor cues.

In the field of anxiety, assessment of behavior in the elevated plus-maze has become one of the most popular and reliable animal models. The elevated plus-maze is used extensively in behavioral testing to identify anxiety-related behaviors and to assess the effects of anxiolytic and anxiogenic agents on these behaviors (Pellow, Chopin, File, & Briley, 1985). Because the elevated plus-maze has proven to be a reliable and credible measure for the assessment of anxiety-related behaviors, it was selected for use in the present study. Rats that remain in the closed arms of the elevated plus-maze are believed to be exhibiting anxiety-related and fear-motivated behaviors, whereas rats that frequently explore the open arms are not considered to be exhibiting anxiety or fear (Pellow, Chopin, File, & Briley, 1985).

Further, it has been suggested that the reluctance of rats to explore the open arms of the maze is associated with the fear of being in the open and vulnerable to possible predatory attack (Fernandes & File, 1996;

Pellow, Chopin, File, & Briley, 1985; Treit, Menard, & Royan, 1993). High levels of locomotion, frequent entries into the central sector, a significant amount of time spent in the central platform, and a significant amount of time spent in the open arms of the maze are indicative of an animal showing relatively lower levels of anxiety. Conversely, low levels of locomotion, infrequent entries into the central sector, little time spent in the central platform, and little time spent in the open arms of the maze are indicative of an animal exhibiting relatively higher levels of anxiety (Pellow, Chopin, File, & Briley, 1985).

Moreover, anti-anxiety treatments such as VNS should increase the amount of time rats spend in the open arms and the number of entries into the open arms, whereas anxiogenic agents such as AMN should have the opposite effect (i.e., a decrease in the amount of time spent in the open arm as well as the number of entries into the open arms). The three target behaviors (number of entries into the central sector, number of sectors crossed in the open arms, and the amount of time spent in the open arms) selected for statistical analyses exemplify the epitome of anxiety-related behavioral responses motivated by fear-arousing stimuli.

During the behavioral testing sessions in the present study, some behavioral responses were counted, other behavioral responses were timed, and all target behaviors were documented, scored and recorded on videotape. Later, each rat's behavior as recorded on the videotape was scored by a second observer who was blind to the treatment conditions (Britton & Britton, 1981; Markus, 2002). The data from both observers was used to establish inter-rater reliability.

2.9.3. PREDATOR SCENT EXPOSURE TASK

This test was conducted in the open field apparatus. A "hide box" (21 cm x 24 cm x 22 cm) with a door (8 cm x 6 cm) was placed in the middle of one wall of the open field. The door opened toward the center of the open field (see Figure 5).

Prior to behavioral testing, a cat collar made of loosely woven elastic material was worn for approximately three weeks by a domesticated male cat. During behavioral testing, the predator-scented cat collar (on Day Three) or a control (unscented) cat collar (on Days One and Two) was suspended 15 cm above the floor of the apparatus on the side opposite the hide box. The cat collar was suspended from a wooden dowel directly above the intersection of specific sectors enabling us to use the same four sectors during each behavioral testing session and to unambiguously define being in proximity to the cat collar.

Figure 5. Photo of the behavioral testing apparatus set-up used in the open field during the predator scent exposure task.

The same behavioral testing apparatus described above, the same behavioral responses measured and listed next, the same three target behaviors analyzed and mentioned later, and the same behavioral testing procedures outlined below were used in the unpublished doctoral dissertation written by Markus (2002). The number of fecal boli, number of rears, number of leans, number of grooming bouts, number of entries into the hide box, number of exits from the hide box, number of times each rat approached the cat collar, time spent in proximity to the cat collar, time spent in the hide box, time spent against the walls, and the amount of time spent freezing were assessed during 15 minutes of behavioral testing (Dielenberg, Hunt, & McGregor, 2001; Markus,

2002) in the appropriate paradigm. The results of Bell's (2007) research were consistent with Markus' findings.

To begin behavioral testing in the present study, each animal was placed in the center of the open field. During the actual 15-minute behavioral testing session while the animal was in the open field, the number of rears, number of leans, number of grooming bouts, number of entries into the hide box, number of exits from the hide box, and number of times each rat approached the cat collar were assessed using a series of manually controlled counters. Next, while viewing the videotape of the 15-minute behavioral testing session and using four manually controlled stopwatches, the researcher assessed time spent in proximity to the cat collar, time spent in the hide box, time spent against the walls, and time spent freezing. At the end of each session, the number of fecal boli was counted via direct observation by the researcher who conducted the behavioral testing session.

Therefore, during the 15-minute sessions, each animal's behavior was observed by the researcher, documented on a data collection sheet, and recorded on videotape. Also, each 15-minute behavioral testing period was timed using a battery-operated timer. Between trials, the open field was swabbed with a 20% ethanol solution to attenuate odor cues.

In addition to the open field and elevated plus-maze, another reliable and credible method of assessing anxiety-related behavioral responses of laboratory rats is to expose them to a natural predator-related stimulus. Thus, the predator scent exposure task is now one of the most popular procedures used in animal research on anxiety and fear. Therefore, it was selected to be used in the present study. As previously mentioned, an influential movement in the study of anxiety has been a shift towards the use of animal models that have greater ethological relevance. This approach attempts to study anxiety by exposing laboratory rats to predator-related stimuli (e.g., fox scent, cat hair, or an owl call) that closely mimic those seen in the wild rather than the more artificial situations that are often used in the laboratory (Rogers, 1997). It can be argued that the use of such naturalistic stimuli allows for greater precision in the evocation of emotional states and defensive behaviors, thereby leading to improved modeling and analysis of fear and anxiety states.

Typically, during behavioral testing sessions using the predator scent exposure task, rats that appear to be anxious or afraid prefer to spend most of the time in the perimeter clinging to the walls of the open field rather than in the center exploring the predator-scented cat collar. These animals also tend to make frequent entries into the hide box, make few exits from the hide box, make few approaches to the cat collar, and spend a significant amount of time in the hide box. A high percentage of time spent in proximity to the walls, low levels of locomotion, and an increase in freezing behavior are indicators of high levels of anxiety. Conversely, a low percentage of time spent in proximity to the walls, high levels of locomotion, and a decrease in freezing behavior are indicators of low levels of anxiety (Prut & Belzung, 2003). The three target behaviors (number of times rat approached cat collar, time spent in proximity to cat collar, and time spent in the hide box) selected for statistical analyses exemplify the epitome of anxiety-related behavioral responses motivated by fear-arousing stimuli. During the behavioral testing sessions in the present study, some behavioral responses were counted, other behavioral responses were timed, and all target behaviors were documented, scored and recorded on videotape. Later, each animal's behavior as recorded on the videotape was scored by a second observer who was blind to the treatment conditions (Britton & Britton, 1981; Markus, 2002). The data from both observers was used to establish inter-rater reliability.

2.10. DATA AND STATISTICAL ANALYSES

The results of the behavioral testing sessions were analyzed using the Statistical Package for the Social Sciences (SPSS) for Windows version 17.0. A between-subjects multivariate analysis of variance (MANOVA) was used to analyze the anxiety-related behaviors (dependent variables) measured in the open field, elevated plus-maze, and predator scent exposure task (independent variables). In this study, three target behaviors were assessed for each behavioral measure used during data collection. The target behaviors measured in the open field were number of sectors crossed, number of entries into the four central sectors, and time spent in the four central sectors. The target behaviors measured in the elevated plus-maze were number of sectors crossed in the open arms, time spent in the open arms, and time spent in the closed arms.

The target behaviors measured in the predator scent exposure task were number of times rats approached cat collar, time spent in proximity to the cat collar, and time spent in the hide box.

A 2 (stimulation type) x 4 (drug conditions) between-subjects MANOVA was used to test for a significant main effect of stimulation and drug dose. The Bonferroni *t* test on *a priori* pairwise comparisons (i.e., comparing one mean with one other mean) and the Fisher's least significant difference (LSD) test on *post hoc* comparisons were intended to be used for multiple group comparisons if the overall *F* was significant as indicated by the follow-up analysis of variance (ANOVA) results that are automatically calculated by the SPSS program and included in the SPSS output after the presentation of the results of Wilks' *Lambda* from the overall MANOVA.

For this research, prior to data collection, it was predicted that electrically stimulated animals in Group One (VNS and saline, 0.5 ml/kg, s.c.) would exhibit fewer anxiety-related behaviors than stimulated animals in Groups Two (VNS and AMN low dose, 0.1 mg/kg, s.c.), Three (VNS and AMN medium dose, 0.5 mg/kg, s.c.), and Four (VNS and AMN high dose, 1.0 mg/kg, s.c.). Thus, we planned to perform a Bonferroni *t* test on the data collected from these groups to determine whether or not the prediction was accurate. The Bonferroni *t* test is particularly appropriate when researchers want to make only a few of all possible comparisons. In addition, the Bonferroni *t* test was selected because it is specifically designed to control familywise error rate. The Bonferroni *t* is based on what is known as the Bonferroni inequality, which states that the probability of occurrence of one or more events can never exceed the sum of their individual probabilities. For example, when three comparisons are made, each with a probability (alpha = 0.05) of a Type I error, the probability of at least one Type I error can never exceed 0.15 (0.05 x 3 = 0.15). This reduces the occurrence of a Type I error. In essence, the Bonferroni *t* runs a regular *t* test but evaluates the result against a modified critical value of *t* that has been chosen so as to limit the familywise error rate (Howell, 1997).

Fisher's LSD test was selected to be used in the present study because Carmer and Swanson (1973) have shown Fisher's LSD to be the most powerful of the common *post hoc* multiple-comparison procedures. Results of their study indicated that Scheffe's test, Tukey's honestly significant difference (HSD) test, and the Student-Newman-Keuls test, for example, are less appropriate *post hoc* tests than Fisher's LSD with the restriction that the F value is statistically significant (Carmer & Swanson, 1973). A significant F is simply an indication that not all the means are equal. However, it does not indicate which means are different from which other means. Thus, if the result of the F test is statistically significant, multiple *post hoc* comparisons should be run to determine which of the group means differ from one another (Howell, 2002).

3.
RESULTS

A 2 (stimulation conditions) x 4 (drug conditions) between-subjects MANOVA was performed on three target behaviors measured in the open field, in the elevated plus-maze, and in the predator scent exposure task. Follow-up ANOVAs were calculated to evaluate the effect of VNS on these behaviors and to assess the capacity of AMN to attenuate the effects of VNS. The results indicated a statistically significant effect of VNS on all behaviors measured. The high dose of AMN (1.0 mg/kg), and in some cases the medium dose (0.5 mg/kg) of AMN, significantly attenuated the effect of VNS on some of these measures.

3.1. OPEN FIELD RESULTS

As noted in the Methods section, a number of measures were taken during behavioral testing in the open field. However, three target behaviors (the number of lines crossed in the open field, the number of entries into the four central sectors, and the amount of time spent in the four central sectors) were chosen to evaluate the effects of VNS on anxiety-related behavioral responses in the open field (Britton & Britton, 1981; Prut & Belzung, 2003) and the capacity of AMN to attenuate peripheral effects of VNS that might mediate the anti-anxiety effect of stimulation. In each case, mean scores for each target behavior were used to assess behaviors measured on Days One and Two (the two baseline days) and to evaluate behaviors measured after stimulation was delivered on Day Three (test day).

Results of an overall between-subjects MANOVA calculated to examine the effect of group on the three target behaviors revealed no statistically significant differences between each of the experimental groups assessed on baseline Day One (*Lambda* (9, 219) = .964, $p > .05$) and baseline Day Two (*Lambda* (9, 219) = .975, $p > .05$). A between-subjects MANOVA was used to analyze Day Three data to examine the effects of Stimulation (VNS vs. Sham) and Drug (saline, 0.5 ml/kg; AMN low dose, 0.1 mg/kg; AMN medium dose, 0.5 mg/kg; AMN high dose, 1.0 mg/kg) on the three target behaviors. The result from the overall Wilks' *Lambda* test indicated that there was a statistically significant stimulation effect when the three target behaviors were measured in the open field (*Lambda* (3, 86) = .341, $p < .01$).

The results from the follow-up univariate ANOVAs indicated that the mean scores for number of lines crossed were significantly affected by VNS compared to sham stimulation (F (1, 88) = 53.78, $p < .01$). Similarly, there was a significant stimulation effect on the number of entries into the four central sectors of the open field (F (1, 88) = 137.30, $p < .01$) as well as on the amount of time spent in the central sectors (F (1, 88) = 52.78, $p < .01$). Graphs of the mean and standard error scores for target behaviors are presented in Figures 6, 7, and 8.

Mean Scores for the Number of Lines Crossed in the Open Field on Day 3

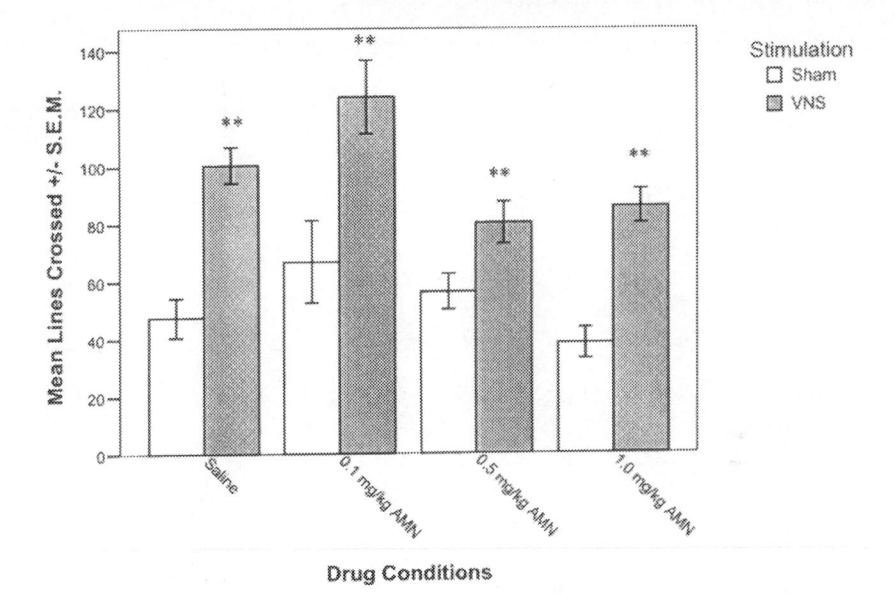

Figure 6. Vagus nerve stimulation significantly increased the number of lines crossed in the open field compared to the saline-treated sham-stimulated animals (** = $p < .01$). There was no statistically significant attenuation of the VNS effect by AMN at any dose compared to the vagus nerve-stimulated saline-treated animals ($p > .05$). Although the drug effect was not statistically significant, the high dose (1.0 mg/kg) of AMN in the sham-stimulated animals appeared to reduce the number of lines crossed ($p > .05$ compared to the saline-treated sham-stimulated animals).

Taunjah P. Bell, Ph.D.

Mean Scores for the Number of Entries into the Four Central Sectors in the Open Field on Day 3

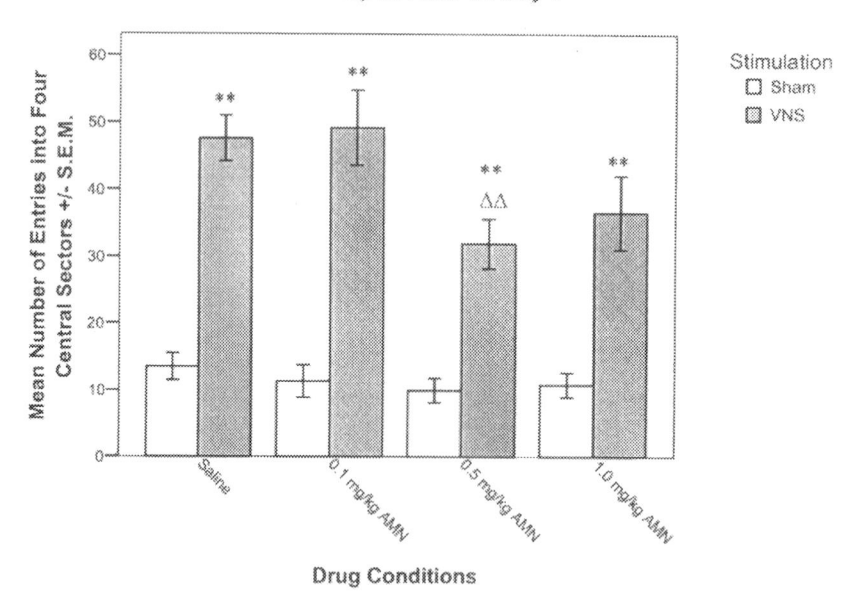

Figure 7. Vagus nerve stimulation significantly increased the number of entries into the four central sectors in the open field during a 15-minute behavioral testing period (** = $p < .01$ compared to the sham-stimulated saline-treated animals). Atropine methyl nitrate at the 0.5 mg/kg dose significantly attenuated the effect of VNS on this measure (ΔΔ = $p < .01$ compared to the vagus nerve-stimulated saline-treated animals).

Mean Scores for the Amount of Time Spent in the Four Central Sectors in Open Field on Day 3

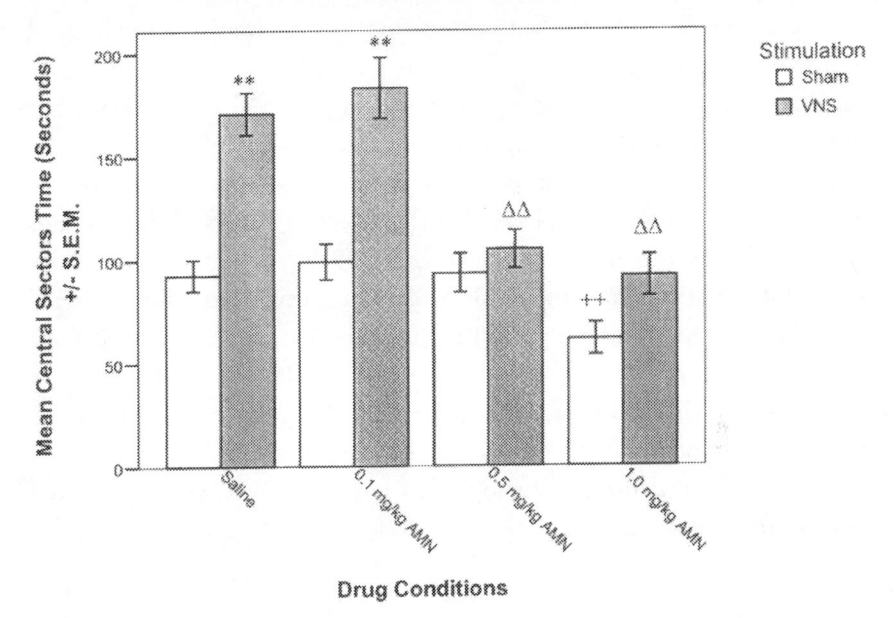

Figure 8. Delivery of VNS 10 minutes prior to behavioral testing resulted in a statistically significant increase in the amount of time spent in the four central sectors only in the saline-treated and AMN low dose-treated animals (** = $p < .01$ compared to the sham-stimulated saline-treated animals). Atropine methyl nitrate at the two higher doses (0.5 mg/kg and 1.0 mg/kg) significantly attenuated the effect of VNS on this measure ($\Delta\Delta = p < .01$ compared to the vagus nerve-stimulated saline-treated animals). The high dose of AMN alone significantly reduced the amount of time spent in the four central sectors (++ = $p < .01$ compared to the saline-treated sham-stimulated animals).

A significant overall drug effect also was found (*Lambda* (9, 209) = .551, $p < .01$) for the three target behaviors measured in the open field. Follow-up univariate ANOVAs indicated that the mean scores for the number of lines crossed were significantly affected by AMN compared to saline (F (3, 88) = 5.54, $p < .05$). Similarly, there was a significant drug effect on the number of entries into the four central sectors of the open field (F (3, 88) = 3.59, $p < .05$) as well as on the amount of time spent in the central sectors (F (3, 88) = 17.94, $p < .01$). Graphs of the mean and standard error scores for each target behavior are presented in Figures 6, 7, and 8.

Post hoc comparisons of individual group means indicated that AMN in the sham-stimulated animals appeared to reduce the number of lines crossed at the high dose (1.0 mg/kg). However, the effect of the high dose of AMN alone on this behavior was not statistically significant ($p > .05$ compared to the saline-treated sham-stimulated animals). Although the medium dose (0.5 mg/kg) of AMN in the sham-stimulated animals appeared to reduce slightly the number of entries into the four central sectors, the effect of AMN alone was not statistically significant at this dose ($p > .05$ compared to the saline-treated sham-stimulated animals). Atropine methyl nitrate in the sham-stimulated animals did not significantly reduce the amount of time spent in the four central sectors at the low and medium doses ($p > .05$ compared to the saline-treated sham-stimulated animals). However, the high dose of AMN alone significantly reduced the amount of time spent in the four central sectors ($p < .01$ compared to the saline-treated sham-stimulated animals).

3.2. ELEVATED PLUS-MAZE RESULTS

As in the open field, a number of measures were taken during behavioral testing in the elevated plus-maze and again three target measures (the number of entries into the central sector, the number of lines crossed in the open arms, and the amount of time spent in the open arms) were chosen to evaluate the effects of VNS on anxiety-related behaviors (Fernandes & File, 1996; File, 1997) and the capacity of AMN to attenuate peripheral effects of VNS that might mediate the anti-anxiety effect of stimulation. In each case, mean scores for each target behavior were used to assess behaviors measured on the two baseline days and behaviors measured after stimulation was delivered on Day Three. Results of an overall between-subjects MANOVA calculated to examine the effect of group on the three target behaviors revealed no statistically significant differences between each of the experimental groups assessed on baseline Day One (*Lambda* (9, 219) = .786, $p > .05$) and baseline Day 2 (*Lambda* (9, 219) = .783, $p > .05$). A between-subjects MANOVA also was calculated on Day Three data to examine the effects of Stimulation (VNS vs. Sham) and Drug (saline, AMN low dose, AMN medium dose, and AMN high dose) on the three target behaviors. The result from the overall Wilks' *Lambda* test indicated that there was a statistically significant effect of stimulation on these three target behaviors (*Lambda* (3, 86) = .253, $p < .01$).

The results from the follow-up univariate ANOVAs indicated that the mean scores for the number of entries into the central sector were significantly affected by VNS compared to sham stimulation (F (1, 88) = 56.10, $p < .01$). Similarly, there was a significant stimulation effect on the number of lines crossed in the open arms of the elevated plus-maze (F (1, 88) = 71.86, $p < .01$) as well as on the amount of time spent in the open arms of the maze (F (1, 88) = 163.11, $p < .01$). Graphs of the mean and standard error scores for each target behavior are presented in Figures 9, 10, and 11.

The result from the overall Wilks' *Lambda* test also revealed a statistically significant drug effect (*Lambda* (9, 209) = .363, $p < .01$). Follow-up univariate ANOVAs indicated that the mean scores for the number of entries into the central sector were significantly affected by AMN compared to saline (F (3, 88) = 20.09, $p < .01$). Similarly, there was a significant drug effect on the number of lines crossed in the open arms of the elevated plus-maze (F (3, 88) = 27.04, $p < .01$) as well as on the amount of time spent in the open arms of the elevated plus-maze (F (3, 88) = 3.57, $p < .05$). Graphs of the mean and standard error scores for each target behavior are presented in Figures 9, 10, and 11.

Mean Scores for the Number of Entries into the Central Sector of the Elevated Plus-Maze on Day 3

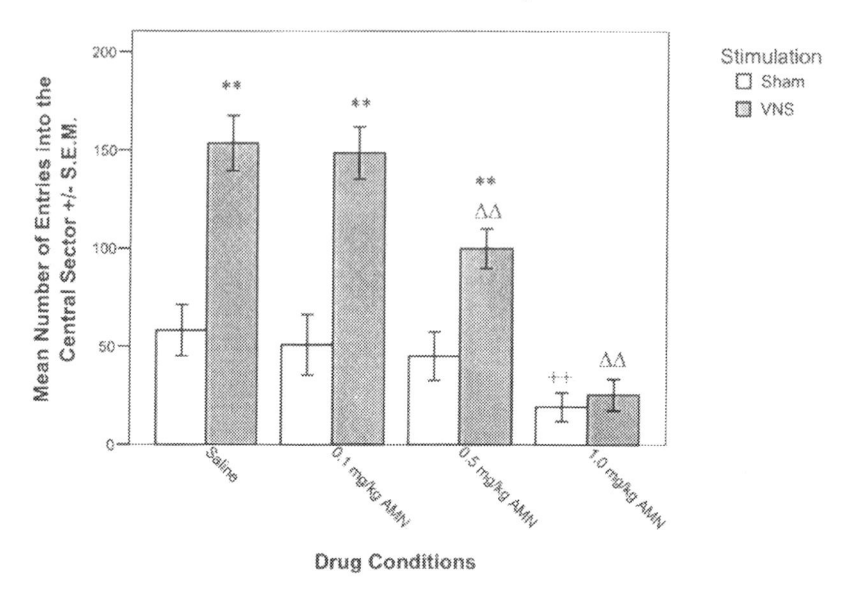

Figure 9. Delivery of VNS 10 minutes prior to the 15-minute testing session resulted in a statistically significant increase in the number of entries into the central sector of the elevated plus-maze in the saline-treated, AMN low dose-treated, and AMN medium dose-treated animals (** = $p < .01$ compared to the sham-stimulated saline-treated animals). Atropine methyl nitrate at the two higher doses (0.5 mg/kg and 1.0 mg/kg) significantly attenuated the effect of VNS on this measure ($\Delta\Delta$ = $p < .01$ compared to the vagus nerve-stimulated saline-treated animals). Atropine methyl nitrate alone significantly reduced the number of entries into the central sector at the high dose (++ = $p < .01$ compared to the saline-treated *sham-stimulated animals*).

Mean Scores for the Number of Lines Crossed in the Open Arms of the Elevated Plus-Maze on Day 3

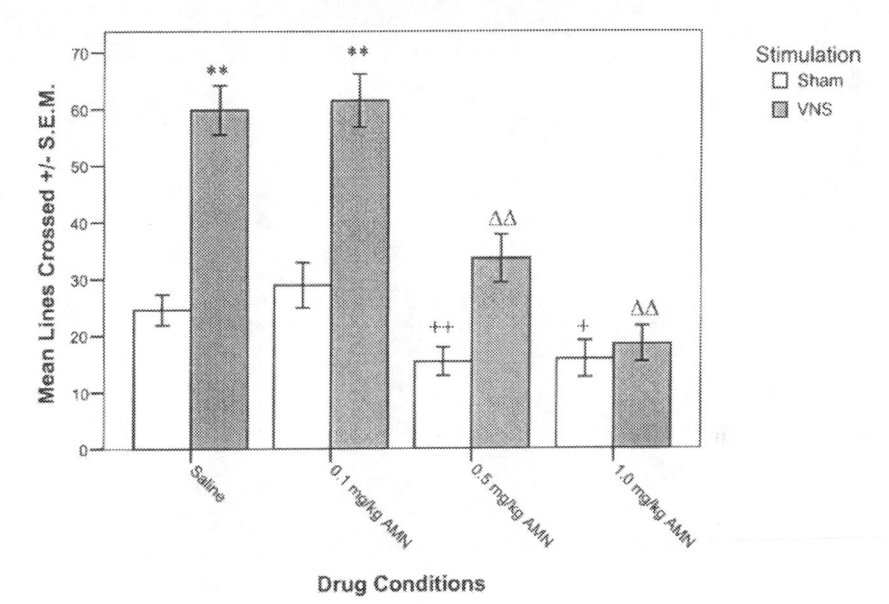

Figure 10. Delivery of VNS 10 minutes prior to the 15-minute testing session resulted in a statistically significantly increase in the number of lines crossed in the saline-treated and AMN low dose-treated animals (** = $p < .01$ compared to the sham-stimulated saline-treated animals). Atropine methyl nitrate at the two higher doses (0.5 mg/kg and 1.0 mg/kg) significantly attenuated the effect of VNS on this behavior (ΔΔ = $p < .01$ compared to the vagus nerve-stimulated saline-treated animals). Atropine methyl nitrate in the sham-stimulated animals significantly reduced the number of lines crossed at the medium (++ = $p < .01$) and high (+ = $p < .05$) doses compared to the saline-treated sham-stimulated animals.

Taunjah P. Bell, Ph.D.

Mean Scores for the Amount of Time Spent in the Open Arms of the Elevated Plus-Maze on Day 3

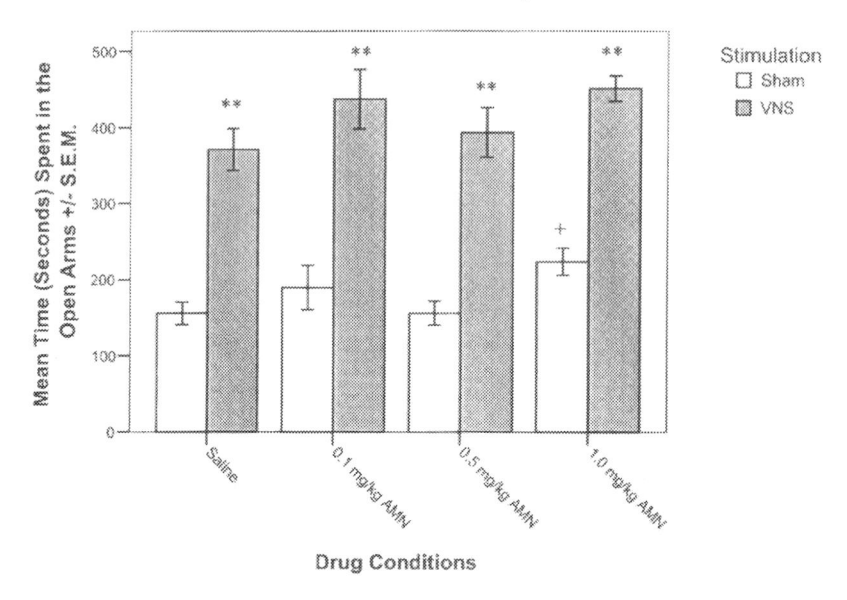

Figure 11. Vagus nerve stimulation significantly increased the amount of time spent in the open arms of the elevated plus-maze during a 15-minute behavioral testing period (** = $p < .01$ compared to the sham-stimulated saline-treated animals). There was no statistically significant attenuation of the VNS effect by AMN at any dose. Interestingly, the high dose (1.0 mg/kg) of AMN in the sham-stimulated animals significantly increased the amount of time spent in the open arms (+ = $p < .05$ compared to the saline-treated sham-stimulated animals). This drug effect is the opposite of the expected effect of the high dose of AMN on this behavior.

Post hoc comparisons indicated that the high dose of AMN in the sham-stimulated animals significantly decreased the number of entries into the central sector of the elevated plus-maze. The medium ($p < .01$) and high ($p < .05$) doses of AMN in the sham-stimulated animals significantly decreased the number of lines crossed in the open arms compared to the saline-treated sham-stimulated animals. Interestingly, the high dose (1.0 mg/kg) of AMN alone significantly increased the amount of time spent in the open arms ($p < .05$ compared to the saline-treated sham-stimulated animals). This result was the opposite of the expected effect of the high dose of AMN on this behavior.

3.3. PREDATOR SCENT EXPOSURE TASK RESULTS

Three target measures (the number of times each rat approached the cat collar, the amount of time spent in proximity to the cat collar, and the amount of time spent in the hide box) also were analyzed for the predator scent exposure task to evaluate the effects of VNS on anxiety-related behaviors (Dielenberg, Arnold, & McGregor, 1999;Dielenberg, Hunt, & McGregor, 2001) and the capacity of AMN to attenuate peripheral effects of VNS. In each case, mean scores for each target behavior were used to assess behaviors measured on the two baseline days and behaviors measured after stimulation was delivered on Day Three. Results of an overall between-subjects MANOVA calculated to examine the effect of group on the three target behaviors revealed no statistically significant differences between each of the experimental groups assessed on baseline Day One (*Lambda* (9, 219) = .815, $p > .05$) and baseline Day Two (*Lambda* (9, 219) = .827, $p > .05$). A between-subjects MANOVA also was calculated on Day Three data to examine the effects of Stimulation (VNS vs. Sham) and Drug (saline, AMN low dose, AMN medium dose, and AMN high dose) on the three target behaviors. A significant stimulation effect was found (*Lambda* (3, 86) = .145, $p < .01$). The result from the overall Wilks' *Lambda* test indicated that there were statistically significant differences between the groups of animals when the three target behaviors were measured during behavioral testing.

The results from the follow-up univariate ANOVAs indicated that the mean scores for the number of times rats approached the predator-scented cat collar were significantly affected by VNS compared to sham stimulation ($F(1, 88) = 209.97, p < .01$). Similarly, there was a significant stimulation effect on the amount of time spent in proximity to the cat collar ($F(1, 88) = 179.87, p < .01$) as well as on the amount of time spent in the hide box ($F(1, 88) = 104.85, p < .01$). Graphs of the mean and standard error scores for each target behavior are presented in Figures 12, 13, and 14.

Taunjah P. Bell, Ph.D.

Mean Scores for the Number of Times Rats Approached the Cat Collar in the Predator Scent Exposure Task on Day 3

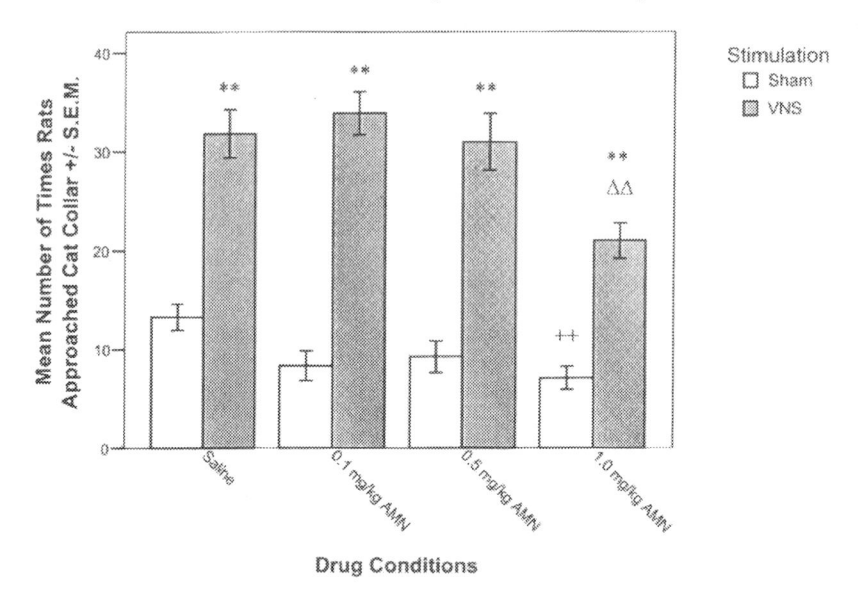

Figure 12. Delivery of VNS 10 minutes prior to the 15-minute testing session resulted in a statistically significant increase in the number of times each rat approached the cat collar in all four drug conditions (saline and all three doses of AMN; ** = $p < .01$ compared to the sham-stimulated saline-treated animals). The high dose of AMN (1.0 mg/kg) significantly attenuated the effect of VNS on this measure (ΔΔ = $p < .01$ compared to the vagus nerve-stimulated saline-treated animals). Atropine methyl nitrate alone significantly reduced the number of times rats approached the cat collar at the high dose (++ = $p < .01$ compared to the saline-treated sham-stimulated animals).

Mean Scores for the Amount of Time Spent in Proximity to the Cat Collar in the Predator Scent Exposure Task on Day 3

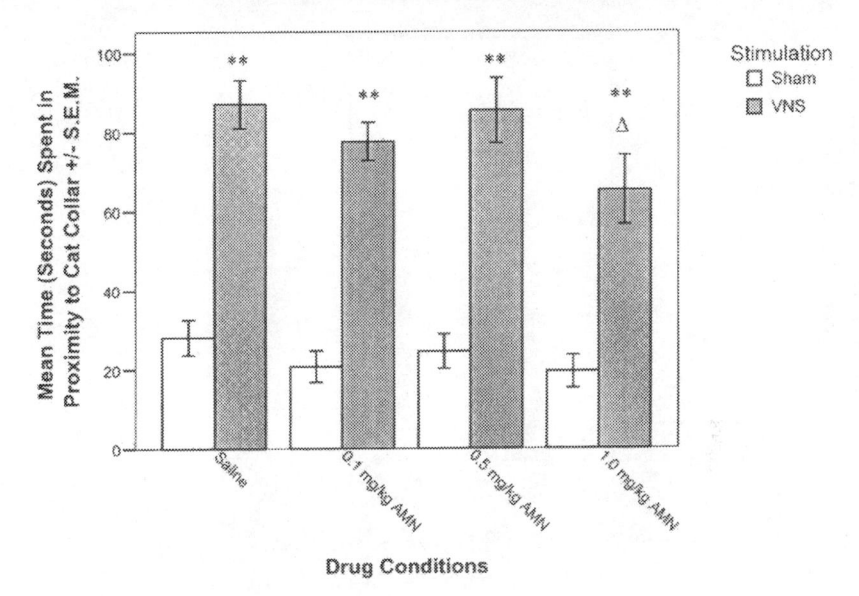

Figure 13. Delivery of VNS 10 minutes prior to the 15-minute testing session resulted in a statistically significant increase in the amount of time spent in proximity to the cat collar in the saline-treated, AMN low-dose-treated, AMN medium dose-treated, and AMN high dose-treated animals (** = $p < .01$ compared to the sham-stimulated saline-treated animals). The high dose of AMN (1.0 mg/kg) significantly attenuated the effect of VNS on this measure ($\Delta = p < .05$ compared to the vagus nerve-stimulated saline-treated animals.

Mean Scores for the Amount of Time Spent in the Hide Box in the Predator Scent Exposure Task on Day 3

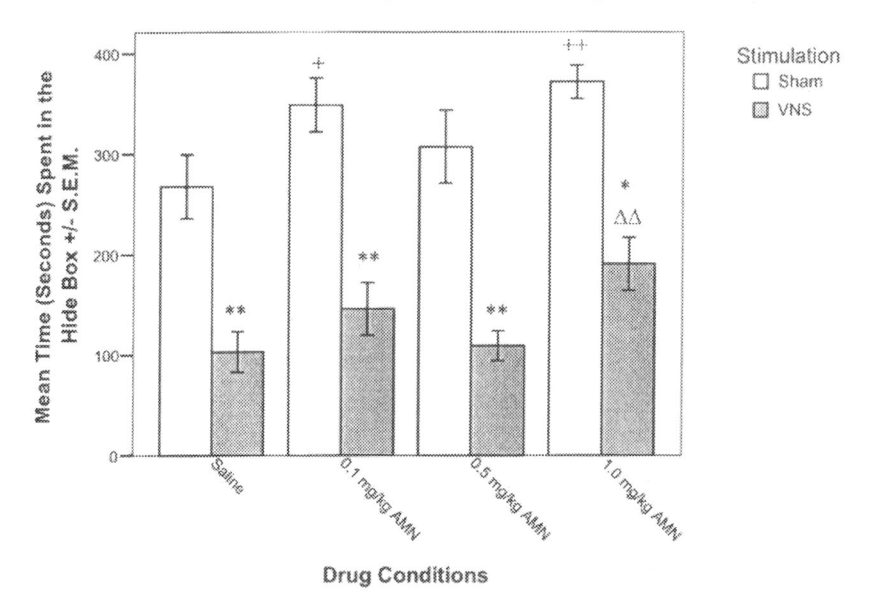

Figure 14. Vagus nerve stimulation significantly decreased the amount of time spent in the hide box in the saline-treated, AMN low dose-treated, AMN medium dose-treated (** = $p < .01$ compared to the sham-stimulated saline-treated animals), and AMN high dose-treated animals (* = $p < .05$ compared to the sham-stimulated saline-treated animals). The high dose of AMN (1.0 mg/kg) significantly attenuated the effect of VNS on this measure (ΔΔ = $p < .01$ compared to the vagus nerve-stimulated saline-treated animals). Atropine methyl nitrate alone significantly increased the amount of time spent in the hide box at the low (+ = $p < .05$ compared to the saline-treated sham-stimulated animals) and high doses (++ = $p < .01$ compared to the saline-treated sham-stimulated animals).

A significant drug effect also was found (*Lambda* (9, 209) = .644, $p < .01$) overall when the three target behaviors were measured during behavioral testing. Follow-up univariate ANOVAs indicated that the mean scores for number of times rats approached cat collar were significantly affected by AMN compared to saline (F (3, 88) = 7.39, $p < .01$). Similarly, there was a significant drug effect on time spent in proximity to the cat collar (F (3, 88) = 2.72, $p < .05$) as well as on time spent in the hide box (F (3, 88) = 5.40, $p < .05$). Graphs of the mean and standard error scores for each target behavior are presented in Figures 12, 13, and 14.

Post hoc comparisons indicated that the low (0.1 mg/kg) and medium (0.5 mg/kg) doses of AMN alone appeared to reduce the number of times rats approached the cat collar. However, the effect of AMN alone at these two doses was not statistically significant ($p > .05$ compared to saline-treated sham-stimulated animals). Atropine methyl nitrate in the sham-stimulated animals appeared to reduce the amount of time spent in proximity to the cat collar at the low and high doses; but, the effect of AMN alone was not statistically significant at these two doses ($p > .05$ compared to saline-treated sham-stimulated animals).

4.
DISCUSSION

4.1. GOALS OF THE PRESENT RESEARCH

It was hypothesized that VNS has the capacity to attenuate anxiety-related behavioral responses. Further, it was posited that VNS might induce its anti-anxiety effects by activating descending parasympathetic efferent fibers thereby producing an increase in parasympathetic tone. If this is indeed the case, then antagonism of cholinergic muscarinic receptors on the target organs with AMN should attenuate the effect of VNS in a dose-related fashion. Thus, it was expected that animals given VNS and saline would exhibit fewer behaviors associated with anxiety than animals given VNS and AMN or sham stimulation and AMN. It also was expected that both stimulated and non-stimulated animals administered the high dose of AMN would exhibit more anxiety-related behavioral responses than animals administered saline or the low dose of AMN. Therefore, the goals of the present study were to determine: (1) whether VNS delivered at the cervical level has the capacity to attenuate behaviors associated with anxiety as measured during performance in the open field, elevated plus-maze, and predator scent exposure task; (2) whether VNS induces its action by activating efferent fibers; (3) whether AMN had some capacity on its own to significantly increase the frequency of behaviors associated with anxiety.

To test the above hypotheses and provide support for future research, the present study was designed to collect data during each animal's 12-

day behavioral testing regimen involving two baseline sessions and one stimulation (VNS or sham) session for each task. During the 15-minute sessions, specific behaviors were measured while animals were tested in the open field for three days, in the elevated plus-maze for three days, and in the predator scent exposure task for three days with a one-day break between each behavioral measure. The nine target behaviors (open field: number of lines crossed, number of entries into the four central sectors, and time spent in the four central sectors; elevated plus-maze: number of entries into the central sector, number of lines crossed in the open arms, and time spent in the open arms; predator scent exposure task: number of times rat approached the cat collar, time spent in proximity to the cat collar, and time spent in the hide box) were the focus of the statistical analyses.

4.2. SUMMARY OF THE FINDINGS

Prior to data collection, animals were assigned to groups based on pre-test locomotor activity ratings to ensure that there was an even distribution of low-, medium-, and high-responders across treatment (stimulation and drug) conditions. Then on Days One and Two of testing in each task (baseline), the animals were acclimated to the open field, elevated-plus maze, and the apparatus used in the predator scent exposure task to further eliminate any pre-existing differences between groups and to confirm that the groups were comparable on each behavior before delivery of VNS. Results of the statistical analyses of baseline data indicated that there were in fact no differences between groups on any of the six baseline test days.

As noted in the results, mean and standard error scores were used to assess the effects of VNS on target behaviors measured after stimulation was delivered on Day Three (test day). In addition, the data were used to determine whether AMN had the capacity to attenuate the effect of VNS in a dose-related fashion. The results indicated that VNS produced a robust and statistically significant effect on all nine target behaviors as well as on some others associated with anxiety. These target behaviors are well accepted as reliable and valid indications of levels of anxiety in rodent models (see Britton & Britton, 1981; Dielenberg, Arnold, & McGregor, 1999; Dielenberg, Hunt, & McGregor, 2001; Fernandes & File, 1996; File, 1997; Hogg, 1996; Pellow, Chopin, File, & Briley, 1985;

Prut & Belzung, 2003). The results provide support for the primary hypothesis that VNS has the capacity to attenuate anxiety.

Further, the findings indicated that in most cases the two higher doses (0.5 mg/kg and 1.0 mg/kg) of AMN had some capacity to attenuate the effect of VNS. The medium dose of AMN significantly reduced the number of entries into the four central sectors of the open field. Both the medium and high doses of AMN significantly reduced the amount of time spent in the four central sectors of the open field. The medium and high doses of AMN also significantly reduced the number of entries into the central sector and the number of lines crossed in the open arms of the elevated plus-maze. In addition, the high dose of AMN had a statistically significant effect on all target behaviors measured in the predator scent exposure task. Atropine methyl nitrate did not significantly reduce the number of lines crossed in the open field or the amount of time spent in the open arms of the elevated plus-maze. Thus, AMN attenuated the effect of VNS on seven of the nine target behaviors. These results are consistent with the expectation that the administration of AMN would attenuate the anxiolytic effects of VNS. Therefore, the goal of determining whether AMN had the capacity to increase significantly anxiety-related behavioral responses was accomplished.

In addition, the results indicated that AMN on its own had some capacity to increase anxiety-related behavioral responses. Rats that received the high dose of AMN coupled with sham stimulation showed more anxiety-related behaviors than saline-treated sham-stimulated animals. For example, in the predator scent task, rats that received the high dose of AMN and no stimulation spent more time in the hide box than did saline-treated sham-stimulated animals. Moreover, animals that received sham stimulation and AMN at the high dose made fewer approaches to the predator-scented cat collar than sham-stimulated saline-treated rats. Further, in the sham-stimulated animals, the high dose of AMN had a significant effect on all target behaviors measured in the elevated plus-maze. Moreover, AMN alone significantly reduced the number of lines crossed in the open arms of the elevated plus-maze at the medium dose compared to saline-treated sham-stimulated animals. In the open field, rats that received the high dose of AMN and no stimulation spent less time in the four central sectors than the saline-treated sham-stimulated animals.

Because AMN on its own had a significant effect on most of the target behaviors measured, the overall statistical analyses revealed that animals administered AMN and no stimulation showed more anxiety-related behaviors during testing than animals administered saline and sham stimulation. Therefore, these results provide evidence to support the hypothesis that AMN on its own had some capacity to affect the target behaviors associated with anxiety in this study. It is not surprising that the results indicated that AMN has anxiogenic properties on its own because antagonism of cholinergic muscarinic receptors results in a reduction in parasympathetic tone (Kottmeier & Gravenstein, 1968; Shutt & Bowes, 1979) causing a relative increase in sympathetic tone resulting in physiological symptoms, such as increases in heart rate and respiration, associated with arousal or anxiety (Hoehn-Saric & McLeod, 1988; Pohjavaara, Telaranta, & Vaisanen, 2003; Porges, McCabe, & Yongue, 1982). This concept is consistent with Porges' (1992, 1995a) theory of autonomic nervous system homeostasis emphasizing the idea that vagal tone is an important component of emotional responsiveness. In response to metabolic demands, that is, for example, an increase in heart rate and respiration. The two components of the autonomic nervous system, the sympathetic and parasympathetic divisions, often function synergistically to yield appropriate responses to the demand, such as modulating cardiovascular output. For example, during exercise there is a progressive decrease in parasympathetic tone and a parallel increase in sympathetic tone (Gellhorn, 1967; Porges, 1992, 1995a).

The autonomic nervous system is not just a response system merely awaiting challenges from the external environment, however. Rather, the autonomic nervous system is continuously modulating visceral afferents in an attempt to promote appropriate physiological states (Gershon, 1998). This regulatory process is mediated primarily by the parasympathetic division. There are, of course, emotional (e.g., stress and anxiety) and physiological states (e.g., fear and arousal) that alter the regulatory function of the autonomic nervous system. These states are characterized by a suppression of parasympathetic tone with a relative excitation of sympathetic tone (Porges, 1992, 1995b). Because the autonomic nervous system is involved in the physiologic expression of stress, shifts in the activity of this system resulting in a disruption of the homeostatic processes seem to characterize the common theme associated with physiologically based definitions of stress (Porges, 1992; Prechtl & Powley, 1990).

Thus, Porges (1995a, 2001) proposed the polyvagal theory to emphasize shifts in the activity of the parasympathetic system, specifically changes in the tone of the vagus. Vagal tone is proposed as a physiologic index of stress, vulnerability to stress, and reactivity to stress. According to Porges' theory, an increase in vagal tone is indicative of a stress-free, homeostatic physiological state whereas a reduction in vagal tone is an indication of stress or a response to stress. Porges' model emphasizes the role of the parasympathetic system and particularly the vagus nerve in mediating physiological states associated with stress. Porges suggests that a measurement of cardiac vagal tone is one method of assessing an individual's response and vulnerability to stress. The method involves monitoring the vagal control of the heart as measured by heart rate variability (Kuwahara, Yayou, Ishii, Hashimoto, Tsubone, & Sugano, 1994) which serves as an index of vagal tone (Levy, 1977). In clinical and experimental settings, this noninvasive measure provides an empirical assessment of the stressful impact of various clinical treatments on patients and aids in the identification of individuals with vulnerabilities to stress (Porges, 1992, 1995a, 1995b, 1998).

The results of the present study suggest that the strong effects of VNS on behavioral measures associated with anxiety produced an increase in vagal tone. This is consistent with Porges' theory in that an increase in vagal tone results in a reduction in peripheral sensations associated with arousal and a decrease in level of anxiety. This also is consistent with James' (1884) solution to the stimulus-to-feeling sequence, which suggests that feedback from physiological events determines feelings. In the case of the present study, the implied increase in vagal tone presumably related to VNS is associated with a perceived reduction in feelings of anxiety and anxiety-related behavioral responses. In addition, the data revealed that AMN had some capacity to reduce the effects of VNS by attenuating the peripheral effects of parasympathetic activation and producing a relative increase in sympathetic tone, a state associated with increased anxiety.

The results of this study also suggest that the implied increase in parasympathetic tone, particularly an increase in vagal tone, produced by VNS can be partially reduced by AMN. Findings indicate that the capacity of AMN to attenuate the effects of VNS is generally significant at the two higher doses. However, because the AMN effects are relatively small compared to the magnitude of the VNS effects, it is likely that there may be a central nervous system (CNS) component associated with the anti-anxiety effects of VNS as well. In other words, both the central and peripheral nervous systems probably contribute to its effects. Moreover, because 70-80% of vagus nerve axons are afferent sensory fibers carrying information to the brain from thoracic and abdominal regions and only 20-30% of vagus nerve axons are efferent motor fibers carrying information from the brain to peripheral cardiac and smooth muscles, organs, and glands, it is highly likely that the ascending vagal afferents to the CNS also play a dominant and important role in mediating the effects of VNS on anxiety.

4.3. DIRECTIONS FOR FUTURE RESEARCH

Further efforts should be made to gain a better understanding of the general effects of VNS on measures of anxiety and its mechanism of action. Markus (2002) suggested that VNS may have caused an overall increase in locomotor activity and thus yielded the appearance of reduced anxiety. However, this is probably not the case because VNS had a significant effect on behaviors that did not involve locomotor activity, such as the amount of time spent in the four central sectors of the open field, amount of time spent in the open arms of the elevated plus-maze, time spent in the hide box in the predator scent exposure task, and importantly the amount of time spent in proximity to the predator-scented cat collar.

Results of research in other laboratories suggest that VNS might induce its effects by facilitating the release of biogenic amines (e.g., serotonin, or 5-HT; dopamine, or DA; norepinephrine, or NE) and this might be associated with the well-known reduction in seizure activity, an attenuation of anxiety, and an improvement in mood. Hammond, Uthman, Wilder, Ben-Menachem, Hamberger, Hedner, et al. (1992) found significant increases in plasma homovanillic acid (a metabolite of DA) and 5-hydroxyindoleacetic acid (a metabolite of 5-HT) in epilepsy patients who received VNS for two months. Plasma levels of these metabolites were measured before the initiation of stimulation (that is, before the stimulator was turned on) and after two months of chronic VNS. The researchers' most salient observations were that the significant increases in these major metabolites also were associated with decreases in seizure frequency and reductions in reports of anxiety combined with improvements in mood.

Because previous research supports the idea that patients suffering from refractory epilepsy often report a reduction in anxiety and an improvement in mood following chronic VNS therapy, it also is plausible that VNS therapy would be an effective treatment in humans for symptoms of anxiety that tend to coexist with depression. Moreover, because pharmacological agents such as anxiolytics (e.g., buspirone, fluoxetine, and propranolol), antidepressants (e.g., amoxapine, clomipramine, and nefazodone), and antiepileptics or anticonvulsants (e.g., lamotrigine, topiramate, tiagabine, valproate, carbamazepine, and gabapentin) have proven to be effective in treating anxiety disorders and clinical depression, the use of VNS (an anticonvulsant agent) for the treatment of anxiety which tends to coexist with depression is currently under investigation (George, Ward, Ninan, Pollack, Nahas, Anderson, et al., 2008; George, Ward, Ninan, Pollack, Nahas, Goodman, et al., 2003). Equally important, because some antiepileptic medications or certain anticonvulsant agents were originally developed to treat anxiety disorders, it is rational to assume that the use of VNS for the treatment of symptoms associated with anxiety also would be effective.

Clinical and non-human animal studies suggest that VNS results in changes in the release of NE (Krahl, Clark, Smith, & Browning, 1998; Roosevelt, Smith, Clough, Jensen, & Browning, 2006), 5-HT (Hammond, Uthman, Wilder, Ben-Menachem, Hamberger, Hedner, et al., 1992), gamma-aminobutyric acid (GABA), and glutamate (Walker, Easton, & Gale, 1999). All are neurotransmitters associated with the pathophysiology of depression and these substances also tend to influence anxiety and related behaviors. In addition, the occurrence of autonomic nervous system dysfunction is well established in patients suffering from major depression (Barnes, Veith, Borson, Verhey, Raskind, & Halter, 1983; Esler, Turbott, Schwarz, Leonard, Bobik, Skews, & Jackman, 1982; Glassman, 1998; Lake, Pickar, Ziegler, Lipper, Slater, & Murphy,

1982; Roy, Pickar, De Jong, Karoum, & Linnoila, 1988; Siever & Davis, 1985; Veith, Lewis, Linares, Barnes, Raskind, Villacres, et al. 1994; Wyatt, Portnoy, Kupfer, Snyder, & Engelman, 1971). George, Sackeim, Marangell, Husain, Nahas, Lisanby, et al. (2000) implied that this autonomic system dysfunction may be a result of a reduction in vagal tone thereby enhancing some of the physiological symptoms associated with depression.

Clinical research has revealed some of the chemical and anatomical systems involved in the etiology of major depression and anxiety disorders (e. g., such as panic disorder, post-traumatic stress disorder, and obsessive-compulsive disorder, or OCD) providing targets for medications to help restore healthy psychological functioning. Research into the neurobiological mechanisms of anxiety in humans has focused on disturbances in CNS noradrenergic functioning (Charney & Heninger, 1986; Uhde, Roy-Byrne, Vittone, Boulenger, & Post, 1985). These studies have linked a reduction in the number of neurons in the locus coeruleus (LC) with the development of some cases of anxiety disorders and clinical depression. The LC projects to a wide variety of structures throughout the brain and it is the primary site of the cell bodies of NE-releasing neurons in the CNS. Another important area of research on the neurobiology of anxiety and depression has focused on the benzodiazepine-GABA receptor system. It has been suggested that the benzodiazepine-GABA system may be neurobiological substrate for generalized anxiety and major depression (Gray, 1985; Insel, 1986). The development of these disorders is associated with an implied reduction in glutamic acid decarboxylase, the major enzyme responsible for the synthesis of GABA.

Researchers have examined the coexistence or overlap between anxiety disorders and depressive illnesses for decades. Data to support this position include the findings that certain anxiety and depressive disorders share common factors. For example, reports of the coexistence of anxiety and depressive states with symptoms of OCD were investigated in the early 1900s (see Janet, 1903). The relationship between OCD and depression is complex, with obsessions appearing to lead to complaints of depression in some cases, obsessions becoming secondary to depression in others, and the phenomena being independent in still others (Rachman & Hodgson, 1980; Sturgis & Meyer, 1980; Welner

& Horowitz, 1975). Indeed, the prevalence of depressive symptoms in this disorder has prompted some researchers to argue that OCD is an affective, rather than an anxiety disorder (Insel, Donnelly, Lalakea, Alterman, & Murphy, 1983). Because anxiety disorders tend to coexist with depressive illnesses, pharmacological agents often prescribed to treat depression (e.g., venlafaxine, paroxetine, sertraline, fluvoxamine, and citalopram) also are effective for treating anxiety. Moreover, because some patients suffering from epilepsy often experience symptoms of anxiety coupled with depression, Alvarez-Silva and colleagues (2006) suggest that these three disorders may share at least some aspects of a common pathogenesis.

Results of animal research have revealed that pertinent and relevant biochemical systems involved in mood are associated with anxiety. This information has facilitated the development of safe and effective drug therapies that more directly target the key neurotransmitters that regulate mood and alleviate the symptoms of depressive disorders that tend to coexist with clinical anxiety. Proponents of the use of VNS for the treatment of drug-resistant depression believe that electrical stimulation of the vagus is an effective and safe way of attenuating the symptoms of this disorder.

If the results of future research indicate that changes in the CNS are mediating some of the anti-anxiety effects of VNS, this will imply that even more studies aimed at understanding the brain structures and/or mechanisms involved should be conducted. Possible methods for studying CNS involvement would include lesions or reversible inactivation of the amygdala. Recall that the amygdala is believed to be critically important in mediating emotional responses (Davis, 1997, 1998a, 1998b) and that this structure receives input from the NTS (Fendt & Fanselow, 1999). The cell bodies of the sensory afferent fibers of the vagus nerve reside in the nodose ganglion and relay information directly to the NTS (Ter Horst & Streefland, 1994). If inactivation of the amygdala decreases the anti-anxiety effects of VNS, this finding will provide strong evidence to support the hypothesis that central processes contribute to the anti-anxiety effects of VNS.

It also would be of interest to conduct an electrophysiological study to gather evidence to examine the effects of VNS on behaviors associated with anxiety and the relevant neurobiological connections from the vagus nerve to the amygdala (Adamec & Morgan, 1994), specifically the basolateral amaygdala or the bed nucleus of the stria terminalis (Davis, 1998b). These structures are believed to be associated with emotion recognition and mood regulation (Van Bockstaele, Peoples, & Valentino, 1999). Perhaps the results of these studies could elucidate the mechanisms involved in the action of VNS on depression which tends to coexist with anxiety. Importantly, studies should be conducted to determine whether the reduction in anxiety-related behaviors following VNS corresponds to an equivalent delay in physiological changes in these or related brain regions.

The results of the present study and research by Markus (2002) raise an interesting question. Is the 10-minute time delay between delivering VNS and conducting behavioral testing optimal? Results of an experiment by Markus (2002) showed that when testing began three minutes following the delivery of stimulation, there was little or no anti-anxiety effect of VNS. However, when a 10-minute delay was used in another experiment by Markus (2002), the results indicated that there were indeed differential effects on anxiety-related behaviors assessed during testing in the open field, elevated plus-maze, and predator scent exposure task. For these reasons, a systematic investigation is warranted to determine the time course of stimulation effectiveness. Perhaps the duration of the anxiolytic effect should be assessed by measuring activity for 30 minutes, divided into five-minute intervals. In the present study, videotapes of the 15-minute testing sessions can be viewed in three segments. Then the target behaviors can be assessed by measuring activity during each five-minute interval to determine whether the anxiety-related behavioral responses were consistently shown across the entire 15-minute session. The results could shed some light on the time course of VNS action. If animals tend to show fewer anxiety-related behaviors during the first five or 10 minutes of testing and differences between the groups become less pronounced during the last five minutes, this would imply that the behavioral effects of VNS tend to dissipate over time.

However, the physiological effects of VNS may continue well beyond the 15-minute test period. For example, heart rate may be affected far longer than behavioral responses. This phenomenon can be tested by measuring the animal's heart rate (Porges, 1992, 1995a) throughout the behavioral testing session and comparing the results to the behavioral data. The results may show that a decreased heart rate is measured following VNS (suggesting an increase in parasympathetic tone) and an increased heart rate is observed following administration of AMN (suggesting an enhancement of anxiety). The results of the behavioral and physiological data may be negatively correlated. This would imply that a decreased heart rate is associated with an increase in vagal tone thereby decreasing anxiety-related physiological sensations whereas an increased heart rate is associated with a decrease in vagal tone thereby increasing anxiety-related behavioral responses.

If the results of reanalyzing the present data in five-minute intervals indicate that animals show more anxiety-related behaviors during the final five minutes of testing, then future behavioral testing sessions should be extended to more accurately assess the effects of VNS. Alternatively, behavioral testing could begin 20, 30, or 40 minutes after VNS is delivered (instead of 10 minutes after VNS is delivered) because the time course of VNS action and its effects might require more time to initiate. If the animals exhibit more anxiety-related behaviors during the last five-minute interval than during the first or second five-minute interval, then it is plausible that a longer delay between the delivery of stimulation and the onset of behavioral testing would be even more effective in producing the anti-anxiety effects of VNS.

4.4. LIMITATIONS OF THE PRESENT RESEARCH

This research project was partially funded by the James Walker Graduate Fellowship Research Award granted to the author by the Graduate School at Southern Illinois University Carbondale with matching funds provided by the Department of Psychology at Southern Illinois University Carbondale. Although these funds facilitated the completion of the present study, more money was needed to purchase additional supplies including telemetry software used to monitor heart rate in small laboratory animals. Due to the high cost of this advanced software and the necessary accessories and equipment (i.e., transmitters, receivers, the data acquisition system, and the data analysis system) required to use this technology, we were unable to monitor heart rate and include measurements of heart rate in the analysis of the data. Therefore, it could not be determined whether heart rate was highly correlated with the anxiety-related behavioral responses measured in this study.

Telemetry systems can be used to monitor physiological signals, such as electrocardiogram (ECG) patterns. For example, in an experiment conducted by Croiset et al. (1994), the effect of a cholinergic blockade by the subcutaneous administration of AMN (0.5 mg/kg) was measured using a telemetry system monitoring electrocardiogram (ECG) patterns. The researchers examined the P-R (interval of the ECG that is a measure of atrioventricular transmission) and R-R (interval of the ECG that is a measure of sinoatrial transmission) intervals of six unrestrained, freely moving rats. Croiset et al. found that the blockade of the parasympathetic system resulted in an increase in heart rate, or P-R interval, and a decrease in the R-R interval. From this and other experiments, the researchers concluded that ECG analysis in freely moving rats can be used to determine the activity of the autonomic nervous system, specifically activity of the parasympathetic system and vagal tone.

Therefore, as previously mentioned in the section on directions for future research, the goals of the next study could be aimed at determining whether or not animals that received AMN at relatively high doses (0.5 mg/kg or 1.0 mg/kg, s.c.) show more anxiety-related behavioral responses than saline-treated sham-stimulated animals or saline-treated vagus nerve-stimulated animals. The question is can this expected increase in anxiety-related behavioral responses supposedly resulting from a subcutaneous injection of AMN be correlated with an increase in heart rate? Another aim could be to determine whether animals that received VNS show fewer behaviors associated with anxiety and a corresponding slower heart rate than animals administered AMN and no stimulation or saline and no stimulation.

Data on heart rate could provide physiological evidence to support the implication that an increase in parasympathetic tone, specifically an increase in vagal tone produced by the electrical stimulation of the vagus nerve, results in a reduction in behaviors associated with anxiety and a slowing of the heart rate. Further, the data also could provide physiological evidence to support the assumption that a decrease in parasympathetic tone, specifically a reduction in vagal tone associated with a cholinergic blockade produced by subcutaneous injections of

AMN at relatively high doses (e.g., 0.5 or 1.0 mg/kg), results in an increase in sympathetic tone, a subsequent enhancement in anxiety-related behavioral responses, and an elevated heart rate.

5. CONCLUSION

As noted in the introduction, anxiety can be a debilitating and sometimes difficult condition to treat. Behavioral interventions are often used and a variety of drug therapies are available worldwide. However, pharmacological treatments have some adverse side effects that limit their usefulness in certain cases. For this reason, alternative strategies for dealing with severe chronic or pathological anxiety may have some usefulness in a subset of patient populations that do not respond well to current drug therapies. Based on the current findings, it may be possible, using VNS, safely and effectively to treat patients suffering from symptoms of clinical anxiety and to attenuate the debilitating behaviors related to this disorder. By administering VNS therapy, clinicians and physicians can avoid exposing their patients to some of the adverse side effects and potential liabilities of drug therapy.

What is envisioned is the possibility that a vagus nerve stimulating device could be implanted in patients suffering from anxiety disorders and these individuals could self-deliver VNS when perceived anxiety levels rise. Thus, for example, an individual suffering from agoraphobia might be able to self-deliver stimulation before venturing outside the confines of home, or an individual suffering from panic disorder might be able to ward off, or at least attenuate, the severity of panic attacks through self-delivered electrical stimulation. Note that VNS would only be self-delivered at times when anxiety levels are expected to rise, such as before boarding a plane (an airplane, that is) or giving a speech.

Taunjah P. Bell, Ph.D.

An individual could be implanted with a radio controlled vagus nerve stimulator and carry a small handheld controlling device in a pocket. When anxiety is expected or experienced, a button could be pushed and stimulation would be delivered to the vagus nerve resulting in an attenuation of symptoms. In such cases, VNS would only be delivered a few times a day, and the probability of patients experiencing the minimal side effects (hoarseness and sore threat, for example) of this particular treatment would be lowered. Clearly, this presumption is speculative. Nonetheless, the implications of the results of the present study point to a potential practical application of these findings.

In this research endeavor, an attempt was made to begin to understand the mechanism of action involved in the effects of VNS on anxiety. The results of this study and previous research by Markus (2002) shed some light on these issues. It appears as though changes in parasympathetic tone, specifically vagal tone, may mediate some aspects of the anxiolytic effects of VNS and its mechanism of action. However, the role of the CNS in the effects of VNS and its mechanism of action on anxiety remain to be elucidated. To the best of my knowledge, to date, there is little published research that directly tests whether parasympathetic components, specifically the vagus nerve, have the capacity to attenuate behaviors associated with anxiety. Nevertheless, the results of a pilot study of the effects of VNS on treatment-resistant anxiety (George, Ward, Ninan, Pollack, Nahas, Anderson, et al., 2008) and similar research may lead to a better understanding of anxiety and the etiology of anxiety disorders, the most common psychiatric disorder. Despite the prevalence and incidence of anxiety disorders in today's society, current behavioral and pharmacological therapies are only partially effective. Perhaps the results of this line of research will help us to find new safe and effective ways of treating clinical anxiety and related disorders.

REFERENCES

Adamec, R. E., & Morgan, H. D. (1994). The effect of kindling of different nuclei in the left and right amygdala on anxiety in the rat. *Physiology and Behavior, 55,* 1-12.

Alvarez-Silva, S., Alvarez-Rodriquez, J., Perez-Echeverria, M. J., & Alvarez-Silva, I. (2006). Panic and epilepsy. *Anxiety Disorders, 20,* 353-362.

Barnes, R. F., Veith, R. C., Borson, S., Verhey, J., Raskind, M. A., & Halter, J. B. (1983). High levels of plasma catecholamines in dexamethasone-resistant depressed patients. *American Journal of Psychiatry, 140,* 1623-1625.

Beckett, S. R., Duxon, M. S., Aspley, S., & Marsden, C. A. (1997). Central *c-fos* expression following 20 kHz/ultrasound-induced defence behaviour in the rat. *Brain Research Bulletin, 42,* 421-426.

Bell, T. P. (2007). *A study of the effects of vagus nerve stimulation on anxiety in laboratory rats* (Unpublished doctoral dissertation). Southern Illinois University, Carbondale, Illinois.

Blanchard, C. D., & Blanchard, R. J. (1972). Innate and conditioned reactions to threat in rats with amygdaloid lesions. *Journal of Comparative and Physiological Psychology, 81,* 281-290.

Blanchard, R. J., Blanchard, D. C., Weiss, S. M., & Meyer, S. (1990). The effects of ethanol and diazepam on reactions to predatory odors. *Pharmacology Biochemistry and Behavior, 35,* 775-780.

Blatt, S. H., & Takahashi, R. N. (1999). Experimental anxiety and the reinforcing effects of ethanol in rats. *Brazil Journal of Medical and Biological Research, 32,* 457-461.

Briegleb, S. K., Gulley, J. M., Hoover, B. R., & Zahniser, N. R. (2004). Individual differences in cocaine- and amphetamine-induced activation of male Sprague-Dawley rats: Contribution of the dopamine transporter. *Neuropsychopharmacology, 29,* 2168-2179.

Britton, D. R., & Britton, K. T. (1981). A sensitive open field measure of anxiolytic drug activity. *Pharmacology Biochemistry and Behavior, 15,* 577-582.

Brudzynski, S. M., & Chiu, E. M. C. (1995). Behavioural responses of laboratory rats to playback of 22 kHz ultrasonic calls. *Physiology and Behavior, 57,* 1039-1044.

Canteras, N. S., Chiavegatto, S., Ribeiro Do Valle, L. E., & Swanson, L. W. (1997). Severe reduction of rat defensive behavior to a predator by discrete hypothalamic chemical lesions. *Brain Research Bulletin, 44,* 297-305.

Canteras, N. S., & Goto, M. (1999). *Fos*-like immunoreactivity in the periaqueductal gray of rats exposed to a natural predator. *NeuroReport, 10,* 413-418.

Carmer, S. G., & Swanson, M. R. (1973). An evaluation of ten pairwise multiple comparison procedures by Monte Carlo methods. *Journal of the American Statistical Association, 68,* 66-74.

Carrasco, G. A., & Van de Kar, L. D. (2003). Neuroendocrine pharmacology of stress. *European Journal of Pharmacology, 463,* 235-272.

Charney, D. S., & Heninger, G. R. (1986). Abnormal regulation of noradrenergic function in panic disorders. *Archives of General Psychiatry, 43,* 1042-1054.

Clark, K. B., Krahl, S. E., Smith, D. C., & Jensen, R. A. (1995). Post-training unilateral vagal stimulation enhances retention performance in the rat. *Neurobiology of Learning and Memory, 63,* 213-216.

Clark, K. B., Smith, D. C., Hassert, D. L., Browning, R. A., Naritoku, D. K., & Jensen, R. A. (1998). Posttraining electrical stimulation of vagal afferents with concomitant vagal efferent inactivation enhances memory storage processes in the rat. *Neurobiology of Learning and Memory, 70,* 364-373.

Croiset, G., Raats, C. J. I., Nijsen, M. J. A. M., & Wiegant, V. M. (1994). Differential effects of cholinergic and adrenergic agents on P-R and R-R intervals in rat ECG. *Neuroscience Research Communications, 14,* 75-84.

Davis, M. (1997). Neurobiology of fear responses: The role of the amygdala. *Journal of Neuropsychiatry and Clinical Neurosciences, 9,* 382-402.

Davis, M. (1998a). Anatomic and physiologic substrates of emotion in an animal model. *Journal of Clinical Neurophysiology, 15,* 378-387.

Davis, M. (1998b). Are different parts of the extended amygdala involved in fear versus anxiety? *Biological Psychiatry, 44,* 1239-1247.

Davis, M. (2000). The role of the amygdala in conditioned and unconditioned fear and anxiety. In J. P. Aggleton (Ed.), *The amygdala: A functional analysis* (2nd ed., pp. 213-287). New York: Oxford University Press.

Davis, M. (2006). Neural systems involved in fear and anxiety measured with fear-potentiated startle. *American Psychologist, 61*(8), 741-756.

Dielenberg, R. A., Arnold, J. C., & McGregor, I. S. (1999). Low-dose midazolam attenuates predatory odor avoidance in rats. *Pharmacology Biochemistry and Behavior, 62,* 197-201.

Dielenberg, R. A., Hunt, G. E., & McGregor, L. S. (2001). 'When a rat smells a cat': The distribution of fos immunoreactivity in rat brain following exposure to a predatory odor. *Neuroscience, 104,* 1085-1097.

Esler, M., Turbott, J., Schwarz, R., Leonard, P., Bobik, A., Skews, H., & Jackman, G. (1982). The peripheral kinetics of norepinephrine in depressive illness. *Archives of General Psychiatry, 39,* 285-300.

Fendt, M., & Fanselow, M. S. (1999). The neuroanatomical and neurochemical basis of conditioned fear. *Neuroscience and Biobehavioral Review, 23,* 743-760.

Fernandes, C., & File, S. E. (1996). The influence of open arm ledges and maze experience in the elevated plus-maze. *Pharmacology Biochemistry and Behavior, 54,* 31-40.

File, S. E. (1985). Animal models for predicting clinical efficacy of anxiolytic drugs: Social behaviour. *Neuropsychobiology, 13,* 55-62.

File, S. E. (1992). Behavioural detection of anxiolytic action. In J. M. Elliot, D. J. Heal, & C. A. Marsden (Eds.), *Experimental approaches to anxiety and depression* (pp. 25-44). New York: John Wiley.

File, S. E. (1997). Animal measures of anxiety. In J. Crawley, C. Gerfen, R. McKay, M. A. Rogawski, D. Sibley, & P. Skolnick (Eds.), *Current protocols in neuroscience* (pp. 8.3.1-8.3.15). New York: John Wiley.

File, S. E., & Gonzalez, L. E. (1996). Anxiolytic effects in the plus maze of 5-HT1A-receptor ligands in dorsal raphe and ventral hippocampus. *Pharmacology Biochemistry and Behavior, 54,* 123-128.

Gellhorn, E. (1967). *Principles of autonomic-somatic integrations: Physiological basis and psychological and clinical implications* (pp. 43-57). Minneapolis, MN: University of Minnesota Press.

George, M. S., Sackeim, H. A., Marangell, L. B., Husain, M. M., Nahas, Z., Lisanby, S. H., et al. (2000). Vagus nerve stimulation: A potential therapy for resistant depression? *The Psychiatric Clinics of North America, 23,* 757-783.

George, M. S., Ward, H. E., Jr., Ninan, P. T., Pollack, M., Nahas, Z., Anderson, B., et al. (2008). A pilot study of vagus nerve stimulation (VNS) for treatment-resistant anxiety disorders. *Brain Stimulation, 1*(2), 112-121.

George, M. S., Ward, H. E., Ninan, P. T., Pollack, M. H., Nahas, Z., Goodman, W. K., et al. (2003, May). *Open trial of VNS in severe anxiety disorders.* Paper presented at the meeting of the American Psychiatric Association, San Francisco, CA.

Gershon, M. (1998). *The second brain.* New York: Harper Collins.

Glassman, A. H. (1998). Depression, cardiac death, and the central nervous system. *Neuropsychobiology, 37,* 80-83.

Gray, J. A. (1985). Issues in the neuropsychology of anxiety. In A. H. Tuma & J. D. Maser (Eds.), *Anxiety and the anxiety disorders* (pp. 121-143). Hillsdale, NJ: Lawrence Erlbaum.

Gulley, J. M., Hoover, B. R., Larson, G. A., & Zahniser, N. R. (2003). Individual differences in cocaine-induced locomotor activity in rats: Behavioral characteristics, cocaine pharmacokinetics, and the dopamine transporter. *Neuropsychopharmacology, 28,* 2089- 2101.

Hall, C. S. (1934). Emotional behavior in the rat: I. Defecation and urination as measures of individual differences in emotionality. *Journal of Comparative Psychology, 18,* 385-403.

Hammond, E. J., Uthman, B. M., Wilder, B. J., Ben-Menachem, E., Hamberger, A., Hedner, T., et al. (1992). Neurochemical effects of vagus nerve stimulation in humans. *Brain Research, 583,* 300-303.

Hoehn-Saric, R. & McLeod, D. R. (1988). The peripheral sympathetic nervous system: Its role in normal and pathologic anxiety. *Psychiatric Clinics of North America, 11,* 375-386.

Hogg, S. (1996). A review of the validity and variability of the elevated plus-maze as an animal model of anxiety. *Pharmacology Biochemistry and Behavior, 54,* 21-30.

Howell, D. C. (1997). *Statistical methods for psychology* (4th ed., pp. 348-399). Belmont, CA: Duxbury Press.

Howell, D. C. (2002). *Statistical methods for psychology* (5th ed., pp. 369-419). Pacific Grove, CA: Duxbury Thomson Learning.

Insel, T. R. (1986). The neurobiology of anxiety: A tale of two systems. In B. F. Shaw, Z. V. Segal, T. M. Vallis, & F. E. Cashman (Eds.), *Anxiety disorders: Psychological and biological perspectives* (pp. 89-103). New York: Plenum.

Insel, T. R., Donnelly, E. F., Lalakea, M. L., Alterman, I. S., & Murphy, D. L. (1983). Neurological and neuropsychological studies of patients with obsessive-compulsive disorder. *Biological Psychiatry, 18,* 741-751.

James, W. (1884). What is an emotion? *Mind, 9,* 188-205.

Janet, P. (1903). *Les obsessions et la psychasthenia.* Paris: Baillière.

Kandel, E. R. (2000). Disorders of mood: Depression, mania, and anxiety disorders. In E. R. Kandel, J. H. Schwartz, & T. M. Jessell (Eds.), *Principles of neural science* (4th ed., pp. 1220-1224). New York: McGraw-Hill.

Korte, S. M., & De Boer, S. F. (2003). A robust animal model of state anxiety: Fear-potentiated behaviour in the elevated plus-maze. *European Journal of Pharmacology, 463,* 163-175.

Korte, S. M., De Boer, S. F., & Bohus, B. (1999). Fear-potentiation in the elevated plus-maze test depends on stressor controllability and fear conditioning. *Stress, 3*, 27-40.

Kottmeier, C. A., & Gravenstein, J. S. (1968). The parasympathomimetic activity of atropine and atropine methylbromide. *Anesthesiology, 29*, 1125-1133.

Krahl, S. E., Clark, K. B., Smith, D. C., & Browning, R. A. (1998). Locus coeruleus lesions suppress the seizure-attenuating effects of vagus nerve stimulation. *Epilepsia, 39*, 709-714.

Kuwahara, M., Yayou, K., Ishii, K., Hashimoto, S., Tsubone, H., & Sugano, S. (1994). Power spectral analysis of heart rate variability as a new method for assessing autonomic activity in the rat. *Journal of Electrocardiology, 27*, 333-337.

Lake, C. R., Pickar, D., Ziegler, M. G., Lipper, S., Slater, S., & Murphy, D. L. (1982). High plasma NE levels in patients with major affective disorder. *American Journal of Psychiatry, 139*, 1315-1318.

Levy, M. N. (1977). Parasympathetic control of the heart. In W. C. Randall (Ed.), *Neural regulation of the heart* (pp. 46-59). New York: Oxford University Press.

Li, H.-Y., & Sawchenko, P. E. (1998). Hypothalamic effector neurons and extended circuitries activated in 'neurogenic' stress: A comparison of footshock effects exerted acutely, chronically, and in animals with controlled glucocorticoid levels. *Journal of Comparative Neurology, 393*, 244-266.

Lin, W. J., Wang, W. W., & Shao, F. (2003). New animal model of emotional stress: Behavioral, neuroendocrine and immunological consequences. *Chinese Science Bulletin, 48*, 1565-1568.

Markus, T. M. (2002). *An investigation into the modulatory effects of vagus nerve stimulation on emotional expression in laboratory rats*. Unpublished doctoral dissertation, Southern Illinois University, Carbondale.

Orr, S. P., Metzger, L. J., & Pitman, R. K. (2002). Psychophysiology of post-traumatic stress disorder. *Psychiatric Clinics of North America, 25*, 271-293.

Ouagazzal, A. M., Kenny, P. J., & File, S. E. (1999). Modulation of behavior on trials 1 and 2 in the elevated plus-maze test of anxiety after systemic and hippocampal administration of nicotine. *Psychopharmacology, 144*, 54-60.

Pellow, S., Chopin, P., File, S. E., & Briley, M. (1985). Validation of open, closed arm entries in an elevated plus-maze as a measure of anxiety in the rat. *Journal of Neuroscience Methods, 14*, 149-167.

Pezzone, M. A., Lee, W.-S., Hoffman, G. E., Pezzone, M. K., & Rabin, B. S. (1993). Activation of brainstem catecholaminergic neurons by conditioned and unconditioned aversive stimuli as revealed by *c-Fos* immunoreactivity. *Brain Research, 608*, 310-318.

Pohjavaara, P., Telaranta, T., & Vaisanen, E. (2003). The role of the sympathetic nervous system in anxiety: Is it possible to relieve anxiety with endoscopic sympathetic block? *Nordic Journal of Psychiatry, 57*, 55-60.

Porges, S. W. (1992). Vagal tone: A physiologic marker of stress vulnerability. *Pediatrics, 90*, 498-504.

Porges, S. W. (1995a). Cardiac vagal tone: A physiological index of stress. *Neuroscience and Biobehavioral Review, 19*, 225-233.

Porges, S. W. (1995b). Orienting in a defensive world: Mammalian modifications of our evolutionary heritage. *Psychophysiology, 32*, 301-318.

Porges, S. W. (1998). Love: An emergent property of the mammalian autonomic nervous system. *Psychoneuroendocrinology, 23*, 837-861.

Porges, S. W. (2001). The polyvagal theory: Phylogenetic substrates of a social nervous system. *International Journal of Psychophysiology, 42,* 123-146.

Porges, S. W., McCabe, P. M., & Yongue, B. G. (1982). Respiratory-heart rate interactions: Psychophysiological implications for pathophysiology and behavior. In J. Cacioppo & R. Petty (Eds.), *Perspectives in cardiovascular psychophysiology* (pp. 223-264). New York: Guilford.

Power, A. E., & McGaugh, J. L. (2002). Cholinergic activation of the basolateral amygdala regulates unlearned freezing behavior in rats. *Behavioural Brain Research, 134,* 307-315.

Prechtl, J. C., & Powley, T. L. (1990). B-afferents: A fundamental division of the nervous system mediating homeostasis. *Behavioral and Brain Sciences, 13,* 289-331.

Prut, L., & Belzung, C. (2003). The open field as a paradigm to measure the effects of drugs on anxiety-like behaviors: A review. *European Journal of Pharmacology, 463,* 3-33.

Rachman, S. J., & Hodgson, R. J. (1980). *Obsessions and compulsions* (pp. 44-56). Englewood Cliffs, NJ: Prentice-Hall.

Rogers, R. J. (1997). Animal models of 'anxiety': Where next? *Behavioral Pharmacology, 8,* 477-496.

Roosevelt, R. W., Smith, D. C., Clough, R. W., Jensen, R. A., & Browning, R. A. (2006). Increased extracellular concentrations of norepinephrine in cortex and hippocampus following vagus nerve stimulation in the rat. *Brain Research, 1119,* 124-132.

Roy, A., Pickar, D., De Jong, J., Karoum, F., & Linnoila, M. (1988). NE and its metabolites in cerebrospinal fluid, plasma, and urine. *Archives of General Psychiatry, 5,* 849-857.

Shutt, L. E., & Bowes, J. B. (1979). Atropine and hyoscine. *Anaesthesia, 34,* 476-490.

Siever, L., & Davis, K. (1985). Overview: Toward a dysregulation hypothesis of depression. *American Journal of Psychiatry, 142,* 1017-1031.

Slaughter, J. S., & Hahn, W. W. (1974). Effects on avoidance performance of vagal stimulation during previous fear conditioning in rats. *Journal of Comparative and Physiological Psychology, 86,* 283-287.

Smith, D. C., Modglin, A. A., Roosevelt, R. W., Neese, S. L., Jensen, R. A., Browning, R. A., et al. (2005). Electrical stimulation of the vagus nerve enhances cognitive and motor recovery following moderate fluid percussion injury in the rat. *Journal of Neurotrauma, 22,* 1485-1502.

Sturgis, E. T., & Meyer, V. (1980). Obsessive-compulsive disorders. In S. M. Turner, K. C. Calhoun, & H. E. Adams (Eds.), *Handbook of clinical behavior therapy* (pp. 68-102). New York: John Wiley.

Tancer, M. E., Lewis, M. H., & Stein, M. B. (1995). Biological aspects. In M. B. Stein (Ed.), *Social phobia, clinical and research perspectives* (pp. 229-257). Washington, DC: American Psychiatric Press.

Ter Horst, G. J., & Streefland, C. (1994). Ascending projections of the solitary tract nucleus. In I. R. A. Barraco (Ed.), *Nucleus of the solitary tract* (pp. 93-103). Boca Raton, FL: CRC Press.

Treit, D., Menard, J., & Royan, C. (1993). Anxiogenic stimuli in the elevated plus-maze. *Pharmacology Biochemistry and Behavior, 44,* 463-469.

Uhde, T. W., Roy-Byrne, P. P., Vittone, B. J., Boulenger, J. P., & Post, R. M. (1985). Phenomenology and neurobiology of panic disorder. In A. H. Tuma & J. D. Laser (Eds.), *Anxiety and the anxiety disorders* (pp. 78-92). Hillsdale, NJ: Lawrence Erlbaum.

Van Bockstaele, E. J., Peoples, J., & Valentino, R. J. (1999). Anatomic basis for differential regulation of the rostrolateral peri-locus coeruleus region by limbic afferents. *Biological Psychiatry, 46,* 1352-1363.

Vazdarjanova, A., Cahill, L., & McGaugh, J. L. (2001). Disrupting basolateral amygdala function impairs unconditioned freezing and avoidance in rats. *European Journal of Neuroscience, 14,* 709-718.

Veith, R. C., Lewis, N., Linares, O. A., Barnes, R. F., Raskind, M. A., Villacres, E. C., et al. (1994). Sympathetic nervous system activity in major depression. *Archives of General Psychiatry, 51,* 411-422.

Walker, B. R., Easton, A., & Gale, K. (1999). Regulation of limbic motor seizures by GABA and glutamate transmission in nucleus tractus solitarius. *Epilepsia, 40,* 1051-1057.

Welner, N., & Horowitz, M. (1975). Intrusive and repetitive thoughts after a depressing experience. *Psychological Reports, 37,* 135-138.

Wyatt, R. J., Portnoy, B., Kupfer, D. J., Snyder, F., & Engelman, K. (1971). Resting plasma catecholamine concentrations in patients with depression and anxiety. *Archives of General Psychiatry, 24,* 65-70.

INDEX

Symbols

5-hydroxyindoleacetic acid 64

A

abdominal 64
acclimated 60
acquisition 18, 69
acrylic 21, 26
action 18, 59, 64, 68, 69, 74, 78
activation 13, 17, 28, 63, 76, 83
active 18
activity 15, 16, 28, 33, 35, 60, 63, 64, 68, 70, 76, 79, 81, 85
acute 17, 28
Adamec, R. E. 75
adhesive 22
administered 8, 23, 26, 29, 30, 31, 59, 62, 70
administration 19, 28, 30, 31, 61, 69, 70, 82
adult 21
affect 62
affective 67, 81
afferent 64, 67
afferents 62, 64, 77, 83, 85
afraid 40
agent 18, 34, 65
agents 36, 37, 65, 67, 77

agoraphobia 73
air 21
airplane 73
alarm 16
AliMed, Inc. 22
alleviate 25, 67
Allied Electronics, Inc. 26
alpha 41
Alterman, I. S. 80
Alvarez-Rodriquez, J. 75
Alvarez-Silva, I. 75
Alvarez-Silva, S. 75
amaygdala 68
ambiguus 14
ammonium 29
amoxapine 65
Amphenol Microminiature Strip Connector 26, 31
amphetamine 28, 76
amplitude 14
A-M Systems, Inc. 21
amygdala 7, 67, 68, 75, 77, 83, 85
amygdalar pathways 7
amygdaloid nuclei 17
analyses 27, 34, 37, 40, 60, 62
analysis 8, 17, 28, 30, 39, 40, 41, 69, 70, 77, 81
analyze 8, 40, 44
anatomical 66

anchor 25
anesthesia 24, 26
animal 7, 14, 15, 16, 21, 23, 24, 25,
    26, 28, 29, 30, 31, 32, 33, 34,
    36, 37, 39, 40, 59, 65, 67, 68,
    77, 80, 81
animals 7, 8, 10, 15, 16, 17, 18, 19,
    21, 25, 27, 28, 29, 30, 40, 41,
    45, 46, 47, 48, 50, 51, 52, 53,
    54, 55, 56, 59, 60, 61, 62, 68,
    69, 70, 81
ANOVA 41
antagonism 19, 59, 62
antagonist 7, 18, 29
anterior 25
anti-anxiety 18, 19, 37, 43, 48, 59, 64,
    67, 68, 69
antibacterial 24
anticonvulsant 65
anticonvulsants 65
antidepressants 65
antiepileptic 65
antiepileptics 65
anxiety 7, 8, 13, 14, 15, 16, 17, 18, 19,
    30, 31, 33, 34, 36, 37, 39, 40,
    41, 43, 48, 53, 59, 60, 61, 62,
    63, 64, 65, 66, 67, 68, 69, 70,
    71, 73, 74, 75, 76, 77, 78, 79,
    80, 82, 83, 84, 85
anxiogenic 14, 15, 16, 17, 18, 34, 36,
    37, 62
anxiolytic 14, 15, 19, 36, 61, 68, 74,
    76, 78
anxiolytics 65
anxious 40
apart 22, 31, 34
apparatus 15, 16, 29, 32, 34, 35, 37,
    38, 60
application 74
applications 14
approach 16, 39
approached 38, 39, 40, 41, 53, 54, 56,
    60
approaches 40, 61, 78

approximately 15, 21, 24, 25, 26, 29,
    30, 34, 37
a priori pairwise comparisons 41
areas 24
arena 15, 16
arms 15, 34, 36, 37, 40, 48, 49, 52, 60,
    61, 64
Arnold, J. C. 78
arousal 14, 62, 63
arousing 34, 37, 40
arrhythmia 14
artificial 16, 39
ascending 64
Aspley, S. 75
assess 14, 26, 36, 43, 48, 53, 60, 69
assessment 14, 36, 63
assessments 30
assigned 17, 27, 28, 60
assignment 28
associated 7, 15, 33, 36, 59, 60, 62, 63,
    64, 65, 66, 67, 69, 70, 74
association 28
atrioventricular transmission 70
atropine methyl nitrate (AMN) 18, 29
atropine sulfate 23
attached 22, 24, 26, 29
attenuate 7, 8, 14, 18, 19, 23, 25, 33,
    36, 39, 43, 48, 53, 59, 60, 61,
    64, 73, 74
author 11, 69
autonomic functions 25
autonomic nervous system dysfunction
    65
autonomic system dysfunction 66
avoidance 17, 18, 78, 84, 85
avoidance performance 18, 84
avoidance responding 18
axons 64

B

Barnes, R. F. 75, 85
baseline 7, 43, 44, 48, 53, 60
basolateral 17, 68, 83, 85
battery-operated timer 33, 36, 39

Beckett, S. R. 75
bedding 21
bed nucleus 68
behavior 14, 15, 16, 17, 18, 31, 32, 33, 34, 35, 36, 37, 39, 40, 43, 47, 48, 49, 51, 52, 53, 56, 60, 76, 79, 82, 83, 84
behavioral 7, 10, 13, 14, 15, 19, 26, 27, 28, 29, 30, 31, 32, 33, 34, 35, 36, 37, 38, 39, 40, 43, 46, 47, 48, 52, 53, 56, 59, 61, 63, 68, 69, 70, 71, 74
behaviors 7, 8, 14, 15, 16, 17, 32, 33, 34, 35, 36, 37, 38, 39, 40, 41, 43, 44, 47, 48, 53, 56, 59, 60, 61, 62, 64, 65, 67, 68, 69, 70, 73, 74, 83
Bell, T. P. 75
Belzung, C. 83
benzodiazepine-GABA 66
between-subjects 8, 40, 41, 43, 44, 48, 53
biochemical 67
biogenic amines 64
biphasic pulse width 7, 26, 30
bipolar stimulating electrode 7
black braided suture material 25
Blanchard, C. D. 75
Blanchard, D. C. 76
Blanchard, R. J. 75, 76
Blatt, S. H. 76
blockade 18, 70
blood-brain barrier 7, 18, 29
blunt dissected 24
Bobik, A. 78
body weight 26
Bohus, B. 81
boli 33, 34, 35, 36, 38, 39
Bonferroni t test 41
Borson, S. 75
Boulenger, J. P. 84
bound 29
bouts 33, 34, 35, 36, 38, 39
Bowes, J. B. 83

box 8, 31, 37, 38, 39, 40, 41, 53, 56, 60, 61, 64
bradycardia 29
brain 7, 15, 17, 18, 29, 64, 66, 67, 68, 78, 79
branch 14
break 29, 31, 60
breathing 14
bregma 25
Briegleb, S. K. 76
bright 33, 34
Briley, M. 82
Britton, D. R. 76
Britton, K. T. 76
Browning, R. A. 77, 81, 83, 84
Brudzynski, S. M. 76
bulb 16
buspirone 65
button 74

C

cage 8, 21, 29, 30, 31
Cahill, L. 85
calvaria 25
Canteras, N. S. 76
capacity 7, 8, 14, 15, 18, 19, 43, 48, 53, 59, 60, 61, 62, 63, 64, 74
carbamazepine 65
Carbondale 10, 11, 21, 22, 69, 75, 82
cardiac 63, 64, 79
cardiovascular 62, 83
caretakers 34
Carmer, S. G. 76
carotid artery 24
Carrasco, G. A. 76
carryover effects 30
cat 10, 16, 17, 30, 31, 37, 38, 39, 40, 41, 53, 54, 55, 56, 60, 61, 64, 78
catecholamines 75
caudal 24
cell bodies 66, 67
cement 26
center 15, 16, 26, 33, 36, 37, 39, 40

central 16, 17, 18, 33, 34, 36, 37, 40, 43, 44, 46, 47, 48, 49, 50, 52, 60, 61, 62, 64, 67, 79
cervical 18, 19, 23, 24, 25, 28, 59
Charles River Laboratories 21
Charney, D. S. 77
chemical 66, 76
Chiavegatto, S. 76
Children's Tylenol, Cherry Flavor 25
Chiu, E. M. C. 76
cholinergic 7, 18, 29, 59, 62, 70, 77
Chopin, P. 82
chronic 65, 73
circuitry 17
circular 15, 16
circulation 26, 29
citalopram 67
Clark, K. B. 77, 81
classified 28
cleanse 24
cleanser 24
clear 17, 21
cling 33
clinging 40
clinical 14, 63, 65, 66, 67, 73, 74, 78, 79, 84
clippers 24
clips 26
clomipramine 65
close 16
closed arms 15, 35, 36, 40
Clough, R. W. 83
CNS 64, 66, 67, 74
cocaine 28, 76, 79
code 27
coexist 65, 67, 68
coexistence 66
cognitive 13, 84
collar 31, 37, 38, 39, 40, 41, 53, 54, 55, 56, 60, 61, 64
collection 26, 30, 33, 36, 39, 40, 41, 60
colony 21
color 16
columns 16

comparisons 41, 42, 48, 52, 57
components 14, 18, 62, 74
compounds 15
concentrations 15, 83, 85
condition 17, 28, 34, 73
conditioned fear 18, 78
conditioned stimulus 18
conditioning 17, 18, 81, 84
conditions 7, 28, 30, 37, 40, 41, 43, 54, 60
conducted 7, 14, 21, 25, 33, 36, 37, 39, 67, 68, 70
connected 26
connections 67
connective 24
connector 24, 26
constructed 21, 22, 23, 26
construction 21
contact 22, 23, 24, 32
contribution 10, 11
control 13, 17, 25, 31, 37, 41, 63, 81
controlled 33, 35, 36, 39, 74, 81
converter 27
Co-Oral-Ite Dental Mfg. 26
corners 33
correlated 69, 70
corticosteroid 15
counted 33, 34, 35, 36, 37, 39, 40
counters 33, 35, 36, 39
coupled 61, 67
course 62, 68, 69
credible 34, 36, 39
Croiset, G. 77
cross 7, 18, 33
crossed 33, 34, 35, 36, 37, 40, 43, 44, 45, 47, 48, 49, 51, 52, 60, 61
cue 16
cues 33, 36, 39
Cyberonics, Inc. 10, 11, 24
Cyberonics Model 101 10
Cyberonics Model 250 26
cycle 21

D

daily 26
data 8, 26, 27, 30, 33, 34, 36, 37, 39, 40, 41, 44, 48, 53, 59, 60, 63, 69, 70
Davis, K. 84
Davis, M. 77
day 29, 30, 31, 43, 60, 74
DC-DC 27
DC-DC converter 27
death 25, 79
debilitating 7, 73
De Boer, S. F. 80, 81
decrease 15, 16, 18, 37, 40, 62, 63, 69, 70
defecation 34
defensive 17, 39, 76, 82
degrees 21, 24
De Jong, J. 83
delay 68, 69
delineated 7
delivered 8, 19, 23, 24, 29, 31, 43, 48, 53, 59, 60, 69, 73
delivery 31, 60, 68, 69
demands 62
dental 26
dependent variables 40
depression 65, 66, 67, 68, 78, 79, 84, 85
depressive 66, 67, 78
deprivation 16
deprived 16
derivative 29
descending 7, 19, 59
design 23
designed 14, 18, 41, 59
detergent 24
development 66, 67
device 73
diagnostic 26
diameter 15, 22, 31
Dielenberg, R. A. 78
difference 10, 41, 42
differences 28, 44, 48, 53, 60, 68, 76, 79
differential 28, 68, 85
direct 33, 36, 39
directions 70
disinfected 24
disorder 7, 66, 67, 73, 74, 80, 81, 82
disorders 13, 65, 66, 67, 73, 74, 77, 79, 80, 84
displacement 24
dissertation 22, 32, 38, 75, 82
dissipate 68
distribution 28, 30, 60, 78
disturbances 66
divisions 62
doctoral 22, 32, 38, 75, 82
documented 33, 34, 36, 37, 39, 40
Donnelly, E. F. 80
dopaminergic transporter function 28
dorsal 17, 24, 25, 78
dorsomedial 24
dose 8, 19, 29, 30, 31, 41, 43, 44, 45, 46, 47, 48, 50, 51, 52, 53, 55, 56, 59, 60, 61, 78
drilled 25
drug 15, 26, 28, 29, 30, 31, 41, 43, 45, 47, 49, 52, 54, 56, 60, 67, 73, 76
drugs 14, 31, 78, 83
dry mouth 13
duration 68
Duxon, M. S. 75

E

Easton, A. 85
effect 8, 15, 19, 29, 37, 41, 43, 44, 45, 46, 47, 48, 49, 50, 51, 52, 53, 54, 55, 56, 59, 60, 61, 62, 64, 68, 70, 75
effectiveness 68
effects 7, 8, 14, 17, 18, 19, 30, 31, 36, 43, 44, 48, 53, 59, 60, 61, 63, 64, 67, 68, 69, 73, 74, 75, 76, 77, 78, 80, 81, 83

efferent 7, 19, 59, 64, 77
electric 24, 31, 34
electrical 7, 17, 23, 24, 25, 67, 70, 73, 77
electrocardiogram (ECG) 70
electrode 22, 23, 24, 25, 26, 28, 32
electrodes 18, 21, 22, 23, 24, 27, 29
electrophysiological 67
elevated 7, 14, 15, 28, 29, 30, 31, 34, 35, 36, 39, 40, 43, 48, 49, 50, 52, 59, 60, 61, 64, 68, 71, 78, 80, 81, 82, 84
elucidate 68
emotion 17, 68, 77, 80
emotional 13, 14, 16, 17, 39, 62, 67, 81
emotionality 15, 16, 79
empirical 63
encircled 22, 23
enclosed 15, 34
Engelman, K. 85
enhancement 69, 71
enhances 77, 84
enhancing 7, 66
entries 15, 16, 33, 34, 35, 36, 37, 38, 39, 40, 43, 44, 46, 47, 48, 49, 50, 52, 60, 61, 82
environment 16, 62
enzyme 66
epilepsy 65, 67, 75
Epoxy 22
equal 28, 30, 35, 42
equipment 10, 14, 69
equivalent 68
Esler, M. 78
ethanol 33, 36, 39, 76
ethical 21
ethological 16, 39
ethologically 17
ethyl alcohol 24
etiology 66, 74
euthanization 26, 32
euthanized 32
events 14, 41, 63
evidence 13, 62, 67, 70

evocation 17, 39
excessive 7
excitation 62
excreted 16
excretion 31, 34
exert 18
exhibit 15, 19, 41, 59, 69
exhibited 17, 33
exhibition 34
exits 38, 39, 40
experience 9, 17, 18, 67, 78, 85
experiment 68, 70
experimental 14, 15, 31, 32, 44, 48, 53, 63
experiments 10, 21, 34, 70
explore 7, 15, 36
exploring 28, 40
expose 22, 24, 39
exposed 17, 22, 23, 24, 28, 34, 76
exposing 16, 39, 73
exposure 7, 10, 14, 17, 18, 29, 30, 31, 33, 37, 38, 39, 40, 41, 43, 53, 59, 60, 61, 64, 68, 78
expression 13, 14, 17, 62, 75, 81
extended 28, 69, 77, 81
external 62
extralaboratory 13

F

F 17, 18, 42, 44, 47, 49, 53, 56, 75, 80, 81, 83, 85
facilitated 67, 69
factors 28, 66
familywise error rate 41
Fanselow, M. S. 78
fashion 19, 59, 60
fat 24
fear 7, 15, 17, 18, 33, 34, 36, 37, 39, 40, 62, 77, 81, 84
fearful 7
fecal 33, 34, 35, 36, 38, 39
feces 16
feedback 63
female 26

Fendt, M. 78
Fernandes, C. 78
fibers 7, 13, 19, 59, 64, 67
field 7, 14, 15, 16, 28, 29, 30, 31, 32, 33, 34, 36, 37, 38, 39, 40, 43, 44, 45, 46, 47, 48, 59, 60, 61, 64, 68, 76, 83
File, S. E. 78, 82
Fine Science Tools 24
Fisher's LSD 41
floor 16, 31, 32, 35, 37
fluoxetine 65
fluvoxamine 67
focus 60
food 16, 21
footshock 8, 16, 17, 31, 34, 81
forceps 24
formation 23
Fos 17, 76, 82
four-hole 26, 31
fox scent 39
frame 25
framework 22, 24
free 21, 63
freely 8, 15, 29, 31, 70
freezing 16, 18, 33, 35, 36, 38, 39, 40, 83, 85
frequency 16, 33, 59, 65
frequent 33, 37, 40
F test 42
function 62, 77, 85
functioning 25, 66
future 59, 64, 67, 69, 70
F value 42

G

gabapentin 65
Gale, K. 85
gastric distress 13
gastric tone 25
Gellhorn, E. 79
generalized 66
generalized anxiety 66
George, M. S. 79

Gershon, M. 79
glands 64
Glassman, A. H. 79
glutamate 65, 85
glutamic acid decarboxylase 66
goal 7, 17, 61
goals 9, 10, 59, 70
gold-plated 26
Gonzalez, L. E. 78
good 22, 23, 32
Goodman, W. K. 79
Goto, M. 76
Gravenstein, J. S. 81
gray 17, 31, 32, 34, 76
Gray, J. A. 79
groom 34
grooming 16, 33, 34, 35, 36, 38, 39
group 17, 18, 29, 30, 41, 42, 44, 48, 53
guidelines 21
Gulley, J. M. 76, 79

H

Hahn, W. W 84
hair 39
half-life 29
Hall, C. S. 79
Halter, J. B. 75
Hamberger, A. 79
Hammond, E. J. 79
handheld 74
Harvard Apparatus 24
Hashimoto, S. 81
Hassert, D. L. 77
HCRs 28
head 8, 24, 25, 26, 29, 31
healing 25, 26
healthy 66
heart 13, 14, 18, 25, 29, 62, 63, 68, 69, 70, 71, 81, 83
heat 24
Hedner, T. 79
height 15
helix 22, 23, 24

Heninger, G. R. 77
Henry Schein, Inc. 23, 26
hide 31, 33, 37, 38, 39, 40, 41, 53, 56, 60, 61, 64
high 8, 15, 16, 25, 28, 29, 30, 32, 34, 40, 41, 43, 44, 45, 47, 48, 50, 51, 52, 53, 54, 55, 56, 59, 60, 61, 69, 70, 71
hoarseness 74
hoc 41, 48, 52, 57
Hodgson, R. J. 83
Hoehn-Saric, R. 80
Hoffman, G. E. 82
Hogg, S. 80
holes 25, 26
home 8, 29, 31, 73
homeostasis 13, 62, 83
homeostatic 63
homovanillic acid 64
honestly 41
Hoover, B. R. 76, 79
horizontal 16, 33
Horowitz, M. 85
hot glass beads 24
hours 16, 26, 29
housed 21
Howell, D. C. 80
HR 28
HSD 42
humans 7, 65, 66, 80
Hunt, G. E. 78
hypothalamic 17, 76
hypothalamus 17
hypothesis 8, 61, 62, 67, 84
hypothesized 19, 59

I

idea 18, 62, 65
Illinois 10, 11, 21, 22, 69, 75, 82
illness 34, 78
illnesses 66, 67
illuminated 15
immunohistochemistry 17
immunoreactivity 17, 76, 78, 82

impedance 26, 27, 32
implantation 23, 25, 28
implanted 7, 18, 22, 23, 25, 27, 29, 31, 73
implications 74, 79, 83
inactivation 67, 77
incision 24, 25, 26
inconsistent 18
increase 7, 15, 16, 17, 29, 37, 40, 47, 50, 51, 54, 55, 59, 61, 62, 63, 64, 69, 70, 71
increased 13, 45, 46, 52, 56, 63, 69
independent 40, 66
index 13, 63, 82
indicating 18
indications 60
indicators 16, 40
individual 10, 14, 28, 41, 48, 63, 73, 79
induce 8, 19, 31, 59, 64
induced 17, 28, 75, 76, 79
inducing 15, 34
induction 14, 33
inescapable 17, 18
infection 25, 26
inhibition 28
injection 7, 23, 26, 29, 70
injections 15, 70
Insel, T. R. 80
inserted 22, 26
instrument 24
instruments 24
insulate 22
intended 29, 41
interpreted 18, 34
inter-rater reliability 34, 37, 40
intersection 24, 38
interval 68, 69, 70
intervals 68, 69, 70, 77
intramuscular (i.m.) 26
intraperitoneal (i.p.) 23
introduction 73
investigation 65, 68, 81
ionization 29
Ishii, K. 81

J

Jackman, G. 78
James, W. 80
James Walker Graduate Fellowship 10, 11, 69
Janet, P. 80
Jensen, R. A. 77, 83, 84

K

Kandel, E. R. 80
Karoum, F. 83
Kenny, P. J. 82
Kopf Model 900 25
Korte, S. M. 80, 81
Kottmeier, C. A. 81
Krahl, S. E. 77, 81
Kupfer, D. J. 85
Kuwahara, M. 81
Kynar insulated 21

L

labeled 16
laboratory 7, 10, 14, 15, 16, 21, 25, 39, 69, 75, 76, 81
Lake, C. R. 81
Lalakea, M. L. 80
Lambda 41, 44, 47, 48, 49, 53, 56
lamotrigine 65
Larson, G. A. 79
lateral 17, 24, 25
laterally 24
layer 24
LCRs 28
lead 27, 66, 74
leaning 16, 33
leans 35, 36, 38, 39
Lee, W.-S. 82
left vagus nerve 24, 25, 29
length 22, 24, 25, 26, 31, 36
Leonard, P. 78
lesions 15, 67, 75, 76, 81
level 14, 18, 19, 24, 25, 26, 59, 63

levels 9, 15, 16, 37, 40, 60, 65, 73, 75, 81
Levy, M. N. 81
Lewis, M. H. 84
Lewis, N. 85
light 16, 21, 68, 74
lighting 16
lights 21, 33, 34
Li, H.-Y. 81
Linares, O. A. 85
lines 16, 32, 33, 34, 43, 44, 45, 47, 48, 49, 51, 52, 60, 61
Linnoila, M. 83
Lin, W. J. 81
Lipper, S. 81
Lisanby, S. H. 79
live 17
locomoted 26
locomotion 15, 16, 24, 28, 33, 37, 40
locomotor 15, 28, 30, 60, 64, 79
locus coeruleus (LC) 66
Long-Evans 7, 21
longitudinally 24
loud 33, 34
low 16, 28, 29, 30, 37, 40, 41, 44, 47, 48, 50, 51, 53, 55, 56, 59, 60
LR 28

M

magnitude 64
major 31, 65, 66, 81, 85
male 7, 21, 26, 37, 76
mammals 13, 14
MANOVA 8, 40, 41, 43, 44, 48, 53
manually 31, 33, 35, 36, 39
Marangell, L. B. 79
Markus, T. M. 81
Marsden, C. A. 75
massaged 26
material 21, 22, 26, 31, 37
Maximum Strength Triple Antibiotic Ointment Plus 25
maze 14, 15, 16, 28, 30, 34, 36, 49, 52, 60, 61, 78

McCabe, P. M. 83
McGaugh, J. L. 83, 85
McGregor, I. S. 78
McGregor, L. S. 78
McLeod, D. R. 80
mean 8, 41, 43, 44, 47, 48, 49, 53, 56, 60
means 42, 48
measure 15, 16, 31, 34, 36, 40, 46, 47, 50, 54, 55, 56, 60, 63, 70, 76, 82, 83
measured 8, 32, 35, 38, 40, 43, 44, 47, 48, 53, 56, 59, 60, 61, 62, 63, 65, 69, 70, 77
measures 7, 19, 30, 43, 48, 53, 63, 64, 78, 79
mechanism 18, 64, 74
medial 17
mediate 43, 48, 74
medications 65, 66
medium 8, 28, 29, 30, 41, 43, 44, 48, 50, 51, 52, 53, 55, 56, 60, 61
memory 7, 77
Menard, J. 84
metabolic 62
metabolite 64
method 14, 15, 39, 63, 81
methods 67, 76, 80
Metzger, L. J. 82
Meyer, V. 84
middle 26, 37
midline 24, 25
mild 8, 31, 34
mimic 16, 39
minimal 74
model 13, 63, 77, 80, 81
modeling 17, 39
models 14, 16, 34, 36, 39, 60, 78, 83
Modglin, A. A. 84
modulating 62
monitor 69, 70
monitored 14, 26, 34
monitoring 13, 14, 63, 70
mood 64, 65, 67, 68, 80
Morgan, H. D. 75

mortality 25
motivated 34, 36, 37, 40
mouse 15
moved 8, 31
movement 16, 39
moving 29, 70
MR 28
multiple 41, 76
multivariate 8, 40
Murphy, D. L. 80, 81
muscarinic 7, 18, 29, 59, 62
muscarinic receptor 7, 18
muscle 22, 24
muscles 24, 64

N

Nahas, Z. 79
naïve 15
Naritoku, D. K. 77
naturalistic 17, 39
NCP Pulse Generator 26
neck 24, 25
nefazodone 65
nerve 7, 10, 11, 13, 14, 17, 18, 21, 22, 23, 24, 25, 26, 27, 28, 31, 45, 46, 47, 50, 51, 52, 54, 55, 56, 63, 64, 67, 68, 70, 73, 74, 75, 79, 80, 81, 83, 84
nervous system 13, 14, 17, 18, 19, 62, 64, 65, 70, 79, 80, 82, 83, 85
neural 13, 17, 80
neurobiological 15, 66, 67
neurobiology 66, 80, 84
NeuroCybernetic Prosthesis (NCP®) 25, 26
neurons 66, 81, 82
neuroscientists 13
neurotransmitters 65, 67
NF 17, 18
Nijsen, M. J. A. M. 77
Ninan, P. T. 79
no 8, 9, 17, 31, 44, 45, 48, 52, 53, 60, 61, 62, 68, 70
nodose ganglion 67

noises 33, 34
noninvasive 14, 63
non-stimulated 59
noradrenergic 66, 77
normal 25, 80
normally 26
notes 25
novel 13
novelty 15
nuclei 75
nucleus 14, 17, 84, 85
number 15, 16, 17, 18, 33, 34, 35, 36, 37, 38, 39, 40, 43, 44, 45, 46, 47, 48, 49, 50, 51, 52, 53, 54, 56, 60, 61, 66
numerical 27

O

objects 16
observation 33, 35, 36, 39
observed 16, 33, 36, 39, 69
obsessions 66
odor 16, 30, 33, 36, 39, 78
open 7, 14, 15, 16, 28, 29, 30, 31, 32, 33, 34, 35, 36, 37, 38, 39, 40, 43, 44, 45, 46, 47, 48, 49, 52, 59, 60, 61, 64, 68, 76, 78, 82, 83
operating temperature 24
opposite 15, 29, 35, 37, 38, 52
optimal 68
organs 18, 19, 59, 64
Orr, S. P. 82
Ouagazzal, A. M. 82
out-bred 28
outer 15
output 41, 62
overactive 13
overall F 41
overlap 66
over-reactive 13
owl call 16, 39

P

PAG 17
pain 25
palpitations 13
panic attacks 73
panic disorder 66, 73, 84
paradigm 39, 83
paradigms 13
parallel 62
parameters 26
parasympathetic 7, 13, 14, 17, 18, 19, 59, 62, 63, 64, 69, 70, 74
paroxetine 67
pathogenesis 67
pathological 73
pathophysiology 65, 83
pathways 14
patients 14, 63, 65, 67, 73, 75, 80, 81, 85
patterns 70
Pellow, S. 82
penetrate 29
penicillin 26
Peoples, J. 85
percentage 15, 16, 40
performance 13, 19, 30, 59, 77
periaqueductal 17, 76
perimeter 33, 40
period 7, 16, 26, 28, 30, 33, 35, 36, 39, 46, 52, 68
periosteum 25
peripheral 18, 43, 48, 53, 63, 64, 78, 80
peripherally-acting 18
periphery 16
Pezzone, M. A. 82
pharmacological 65, 67, 73, 74
phasic 14
phenomena 66
phenomenon 68
physiologic 62, 63, 77, 82
physiological 62, 63, 66, 68, 69, 70, 82
Pickar, D. 81, 83
pins 26

Pitman, R. K. 82
placement 24, 25
plane 73
plasma 15, 29, 64, 75, 81, 83, 85
Plastics One, Inc. 25
plated 22
platform 34, 36, 37
platforms 16
play 17, 64
plus-maze 7, 14, 15, 28, 29, 30, 31, 34, 35, 36, 39, 40, 43, 48, 49, 50, 52, 59, 61, 64, 68, 80, 81, 82, 84
pocket 74
Pohjavaara, P. 82
pole 22, 24
Pollack, M. H. 79
polyvagal theory 63, 83
polyvinyl chloride (PVC) tubes 22
populations 73
Porges, S. W. 82, 83
Portage, Michigan 21
Portnoy, B. 85
posited 19, 59
positioned 25, 26
post 8, 24, 25, 26, 29, 31, 41, 66, 82
posterior 25
postoperatively 25
Post, R. M. 84
Povidone-iodine 24
Power, A. E. 83
Powley, T. L. 83
P-R 70, 77
practical 14, 74
Prechtl, J. C. 83
precision 17, 39
predator 7, 10, 14, 16, 17, 29, 30, 31, 37, 38, 39, 40, 41, 43, 53, 59, 60, 61, 64, 68, 76
predators 33
predatory 16, 36, 76, 78
preparation 22, 25, 29
present 10, 11, 18, 19, 25, 26, 27, 28, 31, 33, 34, 36, 37, 39, 40, 41, 59, 63, 68, 69, 74

presentation 18, 28, 29, 30, 41
presurgery 26
prevent 24, 25, 26
primary 61, 66
principle 7
procedure 23, 24, 26, 30, 31
procedures 16, 24, 32, 35, 38, 39, 41, 76
processes 7, 63, 67, 77
program 26, 41
programming 26
Programming Software 26
progressive 62
promote 25, 26, 62
properties 14, 29, 62
propranolol 65
proteins 29
proximity 16, 38, 39, 40, 41, 53, 55, 56, 60, 64
Prut, L. 83
psychiatric 74
psychological 13, 34, 66, 79
pulsing 24
pure 21

Q

quaternary 29

R

Raats, C. J. I. 77
Rabin, B. S. 82
Rachman, S. J. 83
radio 74
random 28
randomly 17, 29, 31
range 27
rat 15, 16, 23, 24, 25, 26, 28, 29, 37, 38, 39, 40, 53, 54, 60, 75, 76, 77, 78, 79, 81, 82, 83, 84
rate 14, 18, 25, 29, 62, 63, 68, 69, 70, 71, 81, 83
rating 28, 30
ratings 28, 60
rational 65

rats 7, 14, 15, 16, 17, 18, 21, 25, 28, 30, 33, 34, 36, 37, 39, 40, 41, 53, 54, 56, 61, 70, 75, 76, 78, 79, 82, 83, 84, 85
reactions 13, 75, 76
reactivity 63
rearing 16, 33
rears 35, 36, 38, 39
receivers 69
receptor 66, 78
receptors 18, 19, 29, 59, 62
recognition 68
recorded 16, 28, 30, 33, 34, 35, 36, 37, 39, 40
recovered 26
recovery 7, 26, 84
rectangular 16
red 16
reduced 19, 47, 48, 50, 51, 54, 61, 64
reduction 13, 62, 63, 64, 65, 66, 68, 70, 76
reflected 25
reflects 15
refractory 65
regimen 29, 30, 60
region 7, 24, 25, 85
regions 64, 68
regulate 67
regulates 14, 83
regulation 14, 25, 68, 77, 81, 85
regulations 21
regulatory 62
related 7, 13, 14, 15, 19, 34, 36, 37, 39, 40, 41, 43, 48, 53, 59, 60, 61, 62, 63, 65, 68, 69, 70, 71, 73, 74
relay 67
relevance 16, 39
relevant 16, 17, 67
reliable 14, 34, 36, 39, 60
reluctance 15, 36
research 7, 10, 13, 14, 16, 25, 35, 39, 41, 59, 64, 65, 66, 67, 68, 69, 70, 74, 84
Research Award 10, 11, 69

researchers 13, 16, 25, 28, 41, 65, 67, 70
resistant 67, 74, 75, 79
respiration 13, 25, 62
respiratory 14, 23
respiratory problems 23
responders 28, 30, 60
response 7, 14, 15, 28, 30, 62, 63
responses 14, 30, 32, 34, 35, 37, 38, 39, 40, 43, 59, 61, 62, 63, 67, 68, 69, 70, 71, 76, 77
restraint 16
result 13, 17, 25, 41, 42, 44, 48, 49, 52, 53, 66
results 8, 15, 17, 18, 28, 39, 40, 41, 43, 44, 48, 49, 53, 60, 61, 62, 63, 64, 65, 67, 68, 69, 70, 71, 74
retracted 24
reverse 29
reversible 67
Ribeiro Do Valle, L. E. 76
right 17, 25, 75
ring 15
robust 8, 60, 80
rodent 60
rodents 16
Rogers, R. J. 83
role 13, 14, 25, 63, 64, 74, 77, 80, 82
room 21, 24, 35
Roosevelt, R. W. 83, 84
Roy, A. 83
Royan, C. 84
Roy-Byrne, P. P. 84
R-R 70, 77

S

Sackeim, H. A. 79
salient 65
saline 8, 19, 29, 30, 31, 41, 44, 45, 46, 47, 48, 49, 50, 51, 52, 53, 54, 55, 56, 59, 61, 62, 70
salivary gland 24
saturated 29

Sawchenko, P. E. 81
scalp 24, 25
scent 7, 10, 14, 29, 30, 31, 37, 38, 39, 40, 41, 43, 53, 59, 60, 61, 64, 68
Scheffe's 41
Scheffe's test 41
Schwarz, R. 78
scored 30, 31, 34, 37, 40
scores 8, 28, 43, 44, 47, 48, 49, 53, 56, 60
screw 25
screws 26
seams 22
secondary 66
section 22, 26, 43, 70
sections 21
sectors 32, 33, 34, 35, 36, 37, 38, 40, 43, 44, 46, 47, 48, 60, 61, 62, 64
segment 22
segments 23, 68
seizure 64, 65, 81
self-deliver 73
self-delivered 73
sensations 63, 69
sensory 64, 67
separated 22, 24
sequence 63
series 17, 33, 35, 36, 39
sertraline 67
served 22, 24
session 7, 15, 33, 36, 38, 39, 50, 51, 54, 55, 60, 68, 69
settings 13, 63
seven-day 26
sham 8, 29, 30, 31, 44, 45, 46, 47, 48, 49, 50, 51, 52, 53, 54, 55, 56, 59, 61, 62, 70
Shao, F. 81
shape 16
shaved 22, 24
sheets 30
shift 16, 39
shock 8, 17, 18

Shutt, L. E. 83
side 22, 24, 37, 73, 74
Siever, L. 84
Sigma-Aldrich, Corporation 23
signal 17
signals 70
significant 8, 9, 37, 40, 41, 42, 43, 44, 45, 47, 48, 49, 50, 52, 53, 54, 55, 56, 60, 61, 62, 64
signs 34
silver 21, 22, 23, 32
simultaneously 18
sinoatrial transmission 70
sinus 14
site 26, 66
sites 17
situations 16, 39
Skews, H. 78
skull 24, 25, 26, 29
Slater, S. 81
Slaughter, J. S. 84
slowing 29, 70
small 24, 31, 64, 69, 74
Smith, D. C. 77, 81, 83, 84
smooth 64
Snyder, F. 85
sodium chloride 29
sodium pentobarbital (Nembutal) 23
software 26, 69
soldered 21, 22, 26
solution 24, 33, 36, 39, 63
somatic 13, 18, 79
sore threat 74
Southern 10, 11, 21, 22, 69, 75, 82
spaces 15
speech 73
spend 15, 37, 40
spent 15, 16, 28, 33, 34, 35, 36, 37, 38, 39, 40, 43, 44, 47, 48, 49, 52, 53, 55, 56, 60, 61, 64
Sprague-Dawley 28
SS 17, 18
stable 22, 24
staff 21, 32
stainless steel 25, 31

states 13, 17, 39, 41, 62, 63, 66
statistical 27, 34, 37, 40, 60, 62
statistically 42, 43, 44, 45, 47, 48, 49, 50, 51, 52, 53, 54, 55, 57, 60, 61
Statistical Package for the Social Sciences (SPSS) 40
Stein, M. B. 84
step 18
stereotaxic 25
sterilized 24
sterilizer 24
sternohyoid 24
sternomastoid 22, 24
stimulated 8, 18, 23, 41, 45, 46, 47, 48, 50, 51, 52, 54, 55, 56, 59, 61, 70
stimulating 7, 18, 21, 22, 23, 24, 25, 26, 27, 28, 29, 73
stimulation 7, 8, 14, 17, 18, 24, 25, 28, 30, 31, 41, 43, 44, 45, 46, 48, 49, 52, 53, 56, 59, 60, 61, 62, 65, 67, 68, 69, 70, 73, 75, 77, 79, 80, 81, 83, 84
stimulator 65, 74
stimulators 10, 11
stimuli 7, 14, 15, 16, 17, 33, 34, 37, 39, 40, 82, 84
stimulus-specific 7
stimulus-to-feeling 63
stopwatch 28
stopwatches 33, 36, 39
stranded 22, 24
stratified 28
Streefland, C. 84
stress 13, 14, 33, 62, 63, 66, 76, 81, 82
stressful 14, 63
stria terminalis 68
strip 22, 26
structures 66, 67, 68
Student-Newman-Keuls 42
studies 15, 17, 65, 66, 67, 68, 80

study 10, 11, 15, 16, 17, 18, 19, 21, 23, 25, 26, 27, 28, 33, 34, 35, 36, 37, 39, 40, 41, 59, 62, 63, 64, 67, 68, 69, 70, 74, 75, 79
Sturgis, E. T. 84
subcutaneous 7, 23, 24, 29, 70
subcutaneous (s.c.) 23, 29
subject 31, 32
subjects 17, 18, 19, 29, 30, 31, 44, 48, 53
substances 65
substrate 66
substrates 17, 77, 83
sucking 14
Sugano, S. 81
suppressed 13
suppression 62
supraoptic 17
surgery 21, 23, 24, 25, 26, 28, 29, 30
surgical 23, 24, 26
sustained 7
swabbed 24, 33, 36, 39
swallowing 14
Swanson, L. W. 76
Swanson, M. R. 76
swelling 25
sympathetic 13, 17, 62, 63, 71, 80, 82
symptoms 13, 62, 65, 66, 67, 73, 74
synergistically 62
synthesis 66
system 13, 18, 28, 62, 63, 66, 69, 70
systematic 68
systems 64, 66, 67, 70, 77, 80

T

tachycardia 29
Takahashi, R. N. 76
Tancer, M. E. 84
target 8, 18, 19, 32, 34, 35, 37, 38, 40, 43, 44, 47, 48, 49, 53, 56, 59, 60, 61, 62, 67, 68
task 7, 10, 14, 18, 29, 30, 31, 37, 38, 39, 40, 41, 43, 53, 59, 60, 61, 64, 68

technology 69
Telaranta, T. 82
telemetry 69, 70
temperature 21, 24
Ter Horst, G. J. 84
test 15, 16, 28, 29, 30, 31, 35, 37, 41, 43, 44, 48, 49, 53, 59, 60, 68, 81, 82
testing 7, 10, 13, 15, 16, 26, 27, 28, 29, 30, 31, 32, 33, 34, 35, 36, 37, 38, 39, 40, 43, 46, 47, 48, 50, 51, 52, 53, 54, 55, 56, 60, 62, 68, 69
tests 30, 42, 74
theory 62, 63
therapies 67, 73, 74
therapy 65, 73, 79, 84
thigmotaxis 16, 33, 35
thoracic 64
threaded 22, 24
tiagabine 65
time 10, 15, 16, 21, 28, 30, 33, 34, 35, 36, 37, 38, 39, 40, 43, 44, 47, 48, 49, 52, 53, 55, 56, 60, 61, 64, 68, 69
timers 35
tissue 24
tonal 18
tone 7, 13, 14, 17, 18, 59, 62, 63, 64, 66, 69, 70, 71, 74, 82
tonic 14
topiramate 65
total 15, 36
train 7, 26, 30
training 14, 17, 18, 77
translucent 16
transmitters 69
traumatic 66, 82
treated 8, 45, 46, 47, 48, 50, 51, 52, 54, 55, 56, 61, 70
treatment 28, 34, 37, 40, 60, 65, 67, 74, 79
Treit, D. 84
trials 18, 31, 33, 36, 39, 82
trunk 24
tube 22, 24
tubing 22
Tukey's 41
tunnels 16
Turbott, J. 78
Type I error 41

U

Uhde, T. W. 84
ultrasonic 17, 24, 76
underlying 13, 24
univariate 44, 47, 49, 53, 56
University 10, 11, 21, 22, 69, 75, 77, 79, 81, 82
unpredictable danger 7
unrestrained 8, 29, 70
unscented 31, 37
urination 34, 79
urine 16, 31, 83
Uthman, B. M. 79

V

vagal 7, 13, 14, 17, 62, 63, 64, 66, 69, 70, 74, 77, 82, 84
vagus 7, 10, 11, 13, 14, 17, 18, 21, 22, 23, 24, 25, 26, 27, 28, 29, 31, 45, 46, 47, 50, 51, 54, 55, 56, 63, 64, 67, 68, 70, 73, 74, 75, 79, 80, 81, 83, 84
Vaisanen, E. 82
Valentino, R. J. 85
valid 14, 60
valproate 65
value 27, 41
values 27
Van Bockstaele, E. J. 85
Van de Kar, L. D. 76
variability 28, 63, 80, 81
variables 40
variance 8, 40, 41
Vazdarjanova, A. 85
VC 17, 18
Veith, R. C. 75, 85
venlafaxine 67

ventral 24, 25, 78
ventromedial 17
Verhey, J. 75
vertical 16, 33, 35
veterinarians 34
videotape 30, 33, 34, 36, 37, 39, 40
videotaped 28
videotapes 68
Villacres, E. C. 85
visceral 62
Vittone, B. J. 84
Vivarium 21, 32
VNS 7, 8, 14, 18, 19, 23, 29, 30, 31, 37, 41, 43, 44, 45, 46, 47, 48, 49, 50, 51, 52, 53, 54, 55, 59, 60, 61, 63, 64, 65, 67, 68, 69, 70, 73, 74, 79
vocalizations 14, 17
VS 17, 18
vulnerabilities 14, 63
vulnerability 13, 14, 63, 82

W

Wahl Clipper, Corporation 24
Walgreens 22, 24, 25
walk 16
walked 16
Walker, B. R. 85
walls 15, 16, 32, 33, 34, 35, 36, 38, 39, 40
wand 26
Wang, W. W. 81
Ward, H. E. 79
water bottle 25
weekends 26
Weiss, S. M. 76
well-groomed 34
Welner, N. 85
wild 16, 39
Wilder, B. J. 79
Wilks' 41, 44, 48, 49, 53
wire 21, 22, 23, 32
wires 22, 24, 25, 26
wooden dowel 31, 38

wound 26
Wyatt, R. J. 85

Y

Yayou, K. 81
Yongue, B. G. 83

Z

Ziegler, M. G. 81